建筑立场系列丛书 No.90

敬思空间

Transcendental
Architecture

[英]安娜·鲁斯 等 | 编
罗茜 于风军 王方冰 | 译

大连理工大学出版社

敬思空间

建筑立场系列丛书 No.90

004	敬思空间 _ Anna Roos
014	虔诚礼拜堂 _ Mario Filippetto Architetto
024	斯科巴村民活动中心 _ ENOTA
034	法蒂玛圣母礼拜堂 _ Plano Humano Arquitectos
044	水岸佛堂 _ Archstudio
058	Wirmboden高山礼拜堂 _ Innauer-Matt Architekten
070	圣伊利教堂 _ Maroun Lahoud Architecte
078	平衡礼拜堂 _ Álvaro Siza + Carlos Castanheira
090	南美巴哈伊神庙 _ Hariri Pontarini Architects
106	回家 _ Jaap Dawson
112	白俄罗斯纪念礼拜堂 _ Spheron Architects
124	圣温塞斯拉斯教堂 _ Atelier Štěpán
136	阿格里礼拜堂 _ Yu Momoeda Architecture Office
146	苏州礼拜堂 _ Neri&Hu Design and Research Office
160	阿米尔·沙基布·阿斯兰清真寺 _ L.E.FT Architects
172	马里博尔市的唐博斯克教堂 _ Dans Arhitekti
182	塞尔温主教礼拜堂 _ Fearon Hay Architects
194	姆什莱布清真寺 _ John McAslan + Partners
204	莱比锡大学帕琳奈教堂 _ Design Erick van Egeraat
218	建筑师索引

Transcendental Architecture

C3 No.90 Transcendental Architecture

004 Transcendental Architecture_Anna Roos

014 The Votive Chapel_Mario Filippetto Architetto

024 Skorba Village Centre_ENOTA

034 Our Lady of Fatima Chapel_Plano Humano Arquitectos

044 Waterside Buddist Shrine_Archstudio

058 Wirmboden Alpine Chapel_Innauer-Matt Architekten

070 Saint Elie Church_Maroun Lahoud Architecte

078 Chapel of Equilibrium_Álvaro Siza + Carlos Castanheira

090 Baháʼí Temple of South America_Hariri Pontarini Architects

106 Coming Home_Jaap Dawson

112 Belarusian Memorial Chapel_Spheron Architects

124 St. Wenceslas Church_Atelier Štěpán

136 Agri Chapel_Yu Momoeda Architecture Office

146 Suzhou Chapel_Neri&Hu Design and Research Office

160 Amir Shakib Arslan Mosque_L.E.FT Architects

172 Don Bosco Church in Maribor_Dans Arhitekti

182 Bishop Selwyn Chapel_Fearon Hay Architects

194 Msheireb Mosque_John McAslan+Partners

204 Paulinum, Leipzig University_Design Erick van Egeraat

218 Index

敬思空间

Transcendental architecture

是什么造就了一个敬思空间？建筑师如何才能建造出让人产生敬畏感、令人沉思的建筑呢？该如何设计才能让人们消除与建筑物之间的距离感并为之投入情感？

对于敬思空间来说，其本质要求是它既要给使用者带来印象深刻并且超凡的体验，又要是一个提供庇护与安宁之所。只有通过建筑师敏锐的直觉和创造性天赋才能实现这极难把握的要求。敬思空间既复杂又神秘，不仅无法量化敬思空间中的氛围，也无法量化人们的敬拜情感。这使得建筑师的任务更加艰巨，但也更能激发建筑师的兴趣。

What makes a place of worship? How do architects create buildings that foster reverence, contemplation? How can architecture nurture emotional absorption rather than distant observation?

Inherent in the brief for a place of worship is the demand to create a powerful, transcendental experience for users, a place of refuge and peace. This ephemeral demand requires sensibility and creative genius of the architect. Places of worship are complex and mysterious. You can't measure atmosphere and you can't quantify feelings. This makes the task of the architect more demanding, but also more intriguing.

本文将研究建筑师在创造出这些超凡敬思空间时所使用的元素。建筑师是如何对空间、光线、景观和材料进行建模，以创建有助于沉思和祈祷的空间的？无论是黎巴嫩的礼拜堂还是奥地利的高山礼拜堂，智利的神庙或中国的佛堂，卡塔尔的伊斯兰清真寺，还是葡萄牙的童子军礼拜堂，建筑师是如何赋予建筑象征意义的？建筑师又是如何将尘世中的元素升华到神秘的高度？这些错综复杂的问题给建筑师带来了诱人的挑战。

This essay will analyze the elements architects have at their disposal to conjure transcendental places. How have the architects modeled space, light, landscape, and materials to create spaces conducive to contemplation and prayer? How is architecture used symbolically, whether a church in Lebanon or an alpine chapel in Austria, a temple in Chile or a Buddhist shrine in China, an Islamic mosque in Qatar, or a scouts' chapel in Portugal? These complex questions pose an alluring challenge.

敬思空间
Transcendental Architecture

Anna Roos

尽管我并非天主教徒，但最近当我坐在富维耶圣母院教堂中参加弥撒时，我不禁被它的恢弘深深打动。它坐落在里昂的一座山上，是一座新古典主义教堂，柔和的晨光照射在金色的马赛克壁画上，穿白袍的唱诗班男孩高亢的嗓音响彻穹顶，人们一同参与这古老的祈祷仪式，熏香飘荡在整个空间，如此庄严而戏剧化。

自然的力量
我们常在大自然中寻求慰藉，大自然也可以激发我们的情感。日本文化崇尚森林沐浴 (Shinrin-yoku) 这一美好理念，其字面意思是在树木的香脂中洗浴以保持情绪平和。对于建筑师来说，在设计敬思空间时，获得自然的情感力量是非常有意义的。因此，这里所列举的建筑案例中有三分之二的敬思空间都位于自然之中，这并不奇怪。例如，由Archstudio设计的位于中国河北的水岸佛堂（44页）被设计为树下的梯田景观，以符合禅宗佛教与自然的结合的理念。建造这种结构无须砍伐树木，而是将其编织在树木之间，使其真正嵌入景观中。在大自然的包围下，佛教徒踏上通往涅槃的漫长旅程。

本书中许多小礼拜堂都具有与自然相结合的特征。智利的巴哈伊神庙（90页）修建在一个平台上，这个平台通过一段缓缓上升的楼梯与大地相连。神庙坐落在雄伟的安第斯山脉上，像"优雅扭动的翅膀"形成的漩涡，像用蜻蜓精致的翅膀做成的大型豆荚。这座建筑

Although I am not catholic, sitting in the neo-classical Basilica of Notre-Dame de Fourvière perched on a hill above Lyon recently during a catholic mass, I could not help but be moved by the soft morning light reflecting on the golden mosaics, the soprano voice of the white-robed chorister boy ringing high up in the vaults, the congregation all joined together in the ancient ritual of prayer, the tang of incense wafting through the space, the solemnity, the theater.

Force of Nature
We often seek solace in nature and nature can stir our emotions. Japanese culture embraces the wonderful concept of forest bathing (Shinrin-yoku), literally washing yourself in the balm of trees for emotional balance. It makes perfect sense for architects to acquisition the emotive forces of nature when designing places of worship. It is therefore not surprising that two thirds of the places of worship shown here are situated in nature. Archstudio's Waterside Buddhist Shrine in Hebei, China (p.44), for instance, has been designed as a terraced landscape beneath the trees in keeping with Zen Buddhism's call for union with nature. Instead of felling trees, the structure has been woven among the trees and is literally embedded into the landscape. Being enclosed by nature helps Buddhists on their long journey to nirvana.

Many of the chapels featured in this book rejoice in our union with nature. Baháʼí Temple in Chile (p.90) is sited on a plateau accessed by an extended flight of stairs slowly rising up through the landscape. Held by the majestic An-

水岸佛堂，中国
Waterside Buddhist Shrine, China

没有被嵌入景观中，而是被举起来，仿佛是在帮助你达到更好的精神状态。这种有机的形式并不常见，两翼之间半透明的石头和玻璃使其柔和闪亮。

位于奥地利陡峭的Kanisfluh Alp山脚下的Wirmboden高山礼拜堂（58页）是另一个私密的敬思空间的优秀案例，周围环绕着美丽的大自然。这座小礼拜堂位于山坡较低处的山下牧场（当地人称作Vorsäß），是一个举行弥撒的地方，也是传统上农民和他们的牲畜献祭的地方，同时也是一个邻里可以聚会和祈祷的地方。礼拜堂采用当地石材及混凝土建造，在现场收集的石头随意地点缀于夯实的混凝土墙中，就像面包里的葡萄干一样。礼拜堂的形式简洁且谦逊，在传统的棚屋和群山环抱下，这样的形式赋予其很大的精神力量。它是一座乌尔胡特（Urhut），是原始祖先的小屋，所有建筑都是从这样的小屋发展而来的。人们只能钦佩这座小而强大的建筑在其强大的自然环境中的简洁明快，没有丝毫过度的修饰或矫揉造作。

在占地12.1ha的韩国Bugye植物园中，有一座由葡萄牙建筑师阿尔瓦罗·西扎和卡洛斯·卡斯塔涅拉设计的小礼拜堂（78页）。这座纯白色的建筑与深绿色的山脉相映成趣，游客可以在这个避难所里休憩、沉思或祈祷。礼拜堂的外形是一个简单的立方体，外加人字形屋顶，沿着一个倾斜的山脊朝向东方。当太阳升起时，光线通过东墙上的一个小窗慢慢填满礼拜堂，并洒在来这里的人的头上。卡洛斯·卡斯塔涅拉说："这座礼拜堂是一个寻觅神圣与和平的地方，充满着感恩和奇迹。"

dean mountain range, the temple is a whirl of "gracefully torqued wings", like a magnified seedpod made from the delicate wings of a dragonfly. Rather than being embedded in the landscape, this building is held up, as if helping you on your journey to reach a higher mental state. The unusual organic form is softly lit by translucent stone and glazing held between the wings.

Wirmboden Alpine Chapel (p.58) at the foot of the precipitous Kanisfluh Alp in Austria is another excellent example of an intimate place of worship that is surrounded by the raw beauty of nature. Situated on the lower pastures or Vorsäß, the tiny chapel is a place for the celebration of masses and the traditional consecration of farmers and their livestock, a place where neighbors can meet and pray. Stones collected on the site are inserted into tamped concrete walls, like raisins in a bun. Its simplicity and its humble form held in an ensemble of rural sheds and surrounded by mountains give the chapel great spiritual strength. It is an Urhut, the original ancestral hut from which all architecture has descended. One can only admire the lack of excess or pretension in this small, but powerful building in its powerful natural setting.

A small chapel designed by Portuguese architects Álvaro Siza and Carlos Castanheira (p.78) is set amid 12.1 ha of the Bugye arboretum in Korea. It is pure white, striking against the dark green of the mountain. Visitors are able to rest, ponder, or pray for a while, in this shelter. The chapel is oriented eastward along a sloping ridge in the shape of a simple cube and gable roof. When the sun rises, rays of light slowly fill the space through a small window on the eastern wall. Visitors come in and feel the light pouring over their heads. "The chapel is the place to find the Divine

平衡礼拜堂，韩国
Chapel of Equilibrium, Korea

　　在新西兰，位于奥克兰郊区的Parnell住宅区，有一座塞尔温主教礼拜堂（182页），建于圣三一大教堂和圣玛丽教堂之间。这个地方景色宜人，四周橡树成荫。从此处仰望，便可欣赏到奥克兰地标性的火山景观，标志性的一树山就位于其南向轴线上。透过礼拜堂的玻璃，可以看到圣玛丽教堂的维多利亚式屋顶轮廓，其精细的木工制作，还有随季节变化而变化的橡树树冠，实现了"在花园中做礼拜"这一原始设计概念。

　　由Spheron建筑师事务所设计的白俄罗斯纪念礼拜堂（112页）位于伦敦北部的一个花园中。这座礼拜堂犹如一篇关于木材和光线的节奏变换的散文，对于木材和光线的使用恰到好处。白天，礼拜堂看上去很坚固，而到了夜里，它似乎失去了物质性，竟有些虚化，就像树上的灯笼一样散发着微光，照亮外面的世界。它那古雅的洋葱形穹顶，反映了俄罗斯的乡土建筑，让人想起了工业化前的俄罗斯乡村浪漫景象，那个时代在深受人们喜爱的托尔斯泰小说中被永远铭记。就像本书中提到的许多宗教建筑一样，建筑内部允许自然光线的漫射，但往往又不会让朝拜者看到任何外部景色。

　　位于葡萄牙山区的法蒂玛圣母礼拜堂（34页）也与大自然密不可分。这座小礼拜堂坐落于高原之上，俯瞰崎岖的山丘，周围环绕着一片桉树。该礼拜堂最初设计为国家童子军活动营地的避难所，拥有一个简单的折叠式锌屋顶。屋顶的外形就像是一顶折叠帐篷。圣坛之上的礼拜堂屋脊陡然向上，直冲云霄。礼拜堂的各个立面呈梯形，巧妙地立于地面之上，给人一种失重的、神秘的印象。

and Peace with full of Gratitude and Wonder." says Carlos Castanheira.

In Parnell, a residential suburb of Auckland, New Zealand, Bishop Selwyn Chapel (p.182) is built in between Holy Trinity Cathedral and St. Mary's Church. The site is beautiful: the oaks frame an elevated view of the volcanic landmarks of Auckland with the iconic Maungakiekie directly on its southern axis. St. Mary's Victorian roof profile and timber detailing combine with the seasonally changing canopies of the oaks seen through the transparency of the chapel. It fulfills the original concept of "worship in the garden".

The storybook Belarusian Memorial Chapel by Spheron Architects (p.112) is situated in a garden in north London. The chapel is an essay in rhythms of wood and light. During the day the chapel seems solid, while at night it seems to dematerialize, glowing like a lantern in the trees and shining its light to the outside world. With its quaint onion dome, reflecting Russian vernacular architecture, it evokes a romantic vision of pre-industrial rural Russia, a time immortalized in Tolstoy's beloved novels. As with so many of the religious buildings featured in this book, the interior allows in diffuse natural light, while denying worshippers any exterior views.

The chapel dedicated to Our Lady of Fátima in the hills of Portugal (p.34) also seeks communion with nature. Held up on a plateau overlooking the rugged hills and surrounded by a ring of eucalyptus trees, the chapel designed as a shelter for the National Scout Activities Camp, consists of a simple folded zinc roof, a piece of origami reminiscent of a tent. The ridge ascends sharply skyward (heavenward) over the altar. The elevations are trapezoids held delicately above the ground plane, creating an impression of weightlessness or unearthliness.

法蒂玛圣母礼拜堂，葡萄牙
Our Lady of Fátima Chapel, Portugal

宗教象征主义

宗教建筑中总能瞥见象征主义的影子。建筑师会使用各种象征方式，如形状、方位、几何、图案、材料以及光和水。卡塔尔的姆什莱布清真寺（194页）利用水和光取得了极佳的效果。一处长水池落在庭院主入口的轴线上，水池周围由柱廊组成。水也被象征性地用作祈祷之前的净化方式。水的存在给人一种清凉、静默和沉思的感觉，同时水也象征着财富、丰饶和健康。如若说水景常常被用于清真寺外部并赋予其象征意义，那么光则被用于其内部。光束透过重叠并附带图案的网格，在地板和墙面上形成精美的光影。从某种程度上来说，这些光束将有助于将人们"带到"更高的精神领域。由于伊斯兰教禁止在寺庙内出现任何有关阿拉或其他生物的图案，所以在清真寺的装饰上，他们往往选择精美复杂的几何图案。如Jumaa清真寺这样精致的几何图案丰富并增强建筑的整体效果。该设计具有"一个正方形平面不可动摇的几何逻辑，成为和谐统一的象征"，反映了自然界的四种元素：土、气、火和水。可以说该建筑中的每一层、每一处都充满了象征性意义。

与姆什莱布清真寺一样，日本阿格里礼拜堂（136页）的外形也采用正方形平面和立方体体量设计。整座礼拜堂完美地体现了哥特式建筑和经典日式木质体系的巧妙融合，创造了一小片茁壮生长的"森林"，这片"森林"由一根根向上分支的木柱组成。它的整体结构通过支柱垂直得到支撑。在地面上立有四根柱子，在第二层增加到六根柱子，然后像分形花朵一样，增加到八根精致的分支柱子。就像用一

Religious symbolism

Religious architecture is often seeped in symbolism. Architects use a variety of symbolic tools, like form, orientation, geometry, pattern, materials, light and water. Msheireb Mosque in Qatar (p.194) uses water and light to great effect. A long pool of water on axis with the main entrance in the courtyard (sahan) is flanked by colonnades. Water is also used symbolically as a means of purification before prayer, creating a sense of coolness, silence, and contemplation, water also symbolizes wealth, fertility, and health. If water is used to symbolic effect externally, so light is used within the mosque. Pinpoints of light filter through overlaid mesh of cutout patterns, casting delicate shapes onto floor and wall surfaces. These rays of light help to focus attention on higher realms. Islam forbids the representation of Allah or any living creature, so mosques are often adorned with beautifully intricate geometric patterns. Thus, Jumaa mosque is embellished with elaborate geometric shapes that enrich and enhance the architecture. The design has the "unwavering geometrical logic of a square plan, which becomes a symbol of oneness and unity (tawhīd)" and reflects the four pillars (al-arkān) in nature: earth, air, fire, and water. This is architecture imbued with layer upon layer of symbolic meaning.

Like Msheireb Mosque, Agri Chapel (p.136) in Japan also has a square plan and a cubic volume. The chapel is a fascinating marriage of reinterpreted gothic architecture and a quintessential Japanese wooden system to create a soaring forest of timber columns branching heavenward. The structure gains momentum vertically, four columns at ground level, increase to six columns in the second tier, and then, like a fractal bloom, multiplies to eight delicate

苏州礼拜堂，中国
Suzhou Chapel, China

一个简单的算法定义了一个复杂的想法。阿格里礼拜堂有力地诠释了可以运用数学几何解决建筑结构和形式这门艺术，而这也是支撑生命和宇宙的无形逻辑。与姆什莱布清真寺一样，这里的照明也被设计为精致的点点照明，但不同于姆什莱布清真寺，这里的照明光源来自悬挂在木板条天花板上呈矩阵排列的小灯泡。日光透过天花板上方的天窗照进室内，但屋外的景观却因为室内层层饰物的遮挡而变得模糊不清。有趣的是，这样恰好可以防止礼拜者分心，使他能够专注于内心。

　　有时，建筑师会尝试使用巧妙的结构来颠覆建筑形式并回归后现代主义的"滑稽"风格。由Erick van Egeraat建筑设计事务所设计的莱比锡大学新帕琳奈教堂（204页），其设计则是彻底推翻了哥特式建筑的历史传统。教堂的外部由绿松石、灰色玻璃加上石头装饰，而内部乍一看似乎是一个带有精致的罗纹拱顶和支撑圆柱的哥特式风格教堂。然而，这些白色的表面只是薄薄的一层贴花装饰表皮，独立于建筑结构；精美的圆柱也不起支撑作用，而是精美华丽的枝形吊灯灯具，极不协调地悬浮在地板上。阳光透过外立面上细长的拱形窗户漫射进来。窗户的窗棂不是石雕，而是铝材，看起来就像精美的剪纸一般。这样的设计效果可能会令旁观者感到不安和疑惑：建筑师是否试图通过颠覆中世纪建筑的古老传统规范来质疑基本真理？

　　中国苏州礼拜堂（146页）的主体同样也采用了四四方方的设计，一个轻便的立方体体量矗立于由黑砖相互叠放交织所砌成的层层平台基座之上。建筑外墙上一排排金属栅栏如同一层面纱遮挡着里面的实体体量。里面建筑体量的墙上随机地开了许多窗户。在夜晚，

branched columns. Like a simple algorithm that defines a complex idea, Agri Chapel is a powerful essay in the art of creating architectural structure and form with mathematical geometry, which is the invisible logic that underpins life and the universe. As with Msheireb Mosque, here lighting has also been expressed as delicate points, this time in a strict matrix of hanging bulbs and tiny spots attached to the slatted timber ceiling. Daylight filters down through deep skylights above the ceiling and exterior views are obscured by the layering of materials. This lack of exterior views prevents distraction for worshippers, enabling them to focus their minds inwardly.

Sometimes architects try and use clever constructs, subverting forms and reverting to post-modern antics. The new Paulinum of Leipzig University by Design Erick van Egeraat (p.204) has radically subverted the historic norms of gothic architecture. A turquoise and gray glass and stone exterior enclose what at first glance might appear to be a gothic church interior with delicate ribbed vaults and columns. However, the white surfaces are a mere thin appliqué detached from the structure and the elaborate columns do not support the building, but are actually ornate chandelier light fittings hovering incongruously above the floor. Elongated arched windows allow in diffuse light filtered through the exterior facade, The mullions are not carved from stone, but are cut from aluminum, like intricate paper cutouts. The effect is unnerving and somewhat perplexing to the onlooker. Are the architects trying to make one question fundamental truths by subverting the ancient norms of medieval architecture?

Suzhou Chapel (p.146) in China has also been designed on a perfect square plan extruded as a light cubic volume that rises above a layered podium of dark, interwoven brickwork. A perforated metal screen veils the inner solid vol-

圣伊利教堂,黎巴嫩
Saint Elie Church, Lebanon

透过这些窗户,礼拜堂内部的光若隐若现。另外,礼拜堂内的谷仓式屋顶由一根根紧密排列如同音律的木板条组成,这与其外观的颜色和风格截然不同,完全出乎人们的意料。可以说,整座礼拜堂就像是一座藏在盒子里的建筑,是一个远离外部世界的避难所,一个"世外桃源"。

位于黎巴嫩农村地区的圣伊利教堂(70页)的主建筑部分为白色立方体形状,耸立于颇具乡村风格的护围墙体之上。该地区在近代曾遭受暴力战乱而受损严重,而圣伊利教堂正是用来祈求和平的场所。值得注意的是,它是黎巴嫩唯一一座带有钟楼的基督教建筑,钟楼用于召唤群众或祈祷之事,其下方深处的入口就如同一个将世俗世界与神圣空间隔离开的空间。像前面提到过的建筑一样,光线透过窗户进入圣伊利教堂大厅,漫射在位于圣坛后面的墙面上。白色墙壁和光滑的大理石地板上柔和的渐变光线与十字架是这座教堂最精妙的搭配。在这里,纯白色的内饰营造出和平与和谐的氛围。

再来看位于黎巴嫩的另一座清真寺建筑——阿米尔·沙基布·阿斯兰清真寺(160页)。这是一座规模不大的石质清真寺,改建后的清真寺将原有的交叉拱形空间纳为一体。清真寺都应该朝向麦加方向,但是由于原建筑不是朝向麦加方向,因此新清真寺的设计理念就是运用建筑元素,在视觉上改变其朝向,达到朝向麦加的视觉效果。精致的白色宣礼塔横跨于原石质建筑之上,就仿佛半透明的面纱,又好似龙飞凤舞的阿拉伯文字书法。形状颇为抽象的白色宣礼塔由白色钢板条构成。从某个角度看,白塔是钢板一块,厚实坚固,但从另

ume with punctured arrays of random window openings that shine elusively at night. The interior of the chapel, clad in tight rhythms of timber slats with a pitched barn-like roof, completely defies one's expectations. It is a building inside a box, a haven protected from the outer world.

Saint Elie Church (p.70) in a rural district of Lebanon is a white cube rising above a series of rustic retainer walls. A region scarred by violent clashes in its recent history, this church celebrates peace. Interestingly, it is the only Christian project featured here with a bell tower, the traditional call for communal service or prayer. A deep entry beneath the bell tower is a threshold that allows for a transition between the profane outside and the sacred inner world. Again, light is brought indirectly into the main hall, washing against the wall surface behind the altar. The soft gradation of light on white walls and smooth marble floors and the cut out cross are the closest this church gets to adornment. The white interiors instill an atmosphere of peace and harmony.

Also in Lebanon is Amir Shakib Arslan Mosque (p.160), a small-scale stone mosque that incorporates an existing cross-vaulted space. As the existing building is not oriented toward Mecca, as mosques ought, the design concept was to use architectural elements to visually shift the orientation. A delicate white minaret, slung across the solid stone structure is like a translucent veil or the beautiful calligraphic curves of an Arabic letter. Built up of white steel fins, the abstract minaret is at once solid and then seems to dematerialize depending on one's vantage point. Three arches opening onto a public square allow abundant daylight into the domed prayer space. This is an unusually informal, extroverted mosque.

马里博尔市的唐博斯克教堂，斯洛文尼亚
Don Bosco Church in Maribor, Slovenia

圣温塞斯拉斯教堂，捷克共和国
St. Wenceslas Church, Czech Republic

一个角度看，白塔又变得虚无，似乎消失不见了。朝向公共广场有三个拱门，充足的阳光可以透过拱门照进建筑内部带有穹顶的祈祷区域。黎巴嫩的这座清真寺与众不同，不循旧规，不拘礼节，属于外放型。

虔诚礼拜堂位于意大利的科摩（14页），体量小巧，外观是非常规则的三维几何体，其三角的构造就像一块折叠的白色餐巾。阳光通过垂直墙面上的小圆孔照入室内，就像天上的星星，代表着仙后星座和北极星。很神奇的是，从某些角度来看，小礼拜堂似乎是在一个平面的边缘摇摇欲坠，该礼拜堂设计如同姆什莱布清真寺和阿格里礼拜堂的设计，使用了几何结构和严格对称的逻辑。

斯洛文尼亚的唐博斯克教堂（172页）是慈幼会教徒社区一个大型建筑综合体的一部分。整个建筑综合体围绕一个庭院而规划设计。院子里有一棵古老的菩提树。院子四周是廊道，由一些细长精美的柱子支撑着。围绕中心庭院的教堂空间布局遵循了欧洲古代修道院的样式。该建筑综合体的焦点就是弧形的石鼓。石鼓的表面由颜色深浅不同的釉面砖装饰，富有韵律和节奏感，这样的设计使石鼓显得更加柔和。另外，石鼓饰有精致的混凝土王冠，让人联系起基督的荆棘王冠。教堂由混凝土建造，其内部呈圆筒状，上方有一个可以透入自然光的巨大圆形天窗，成为一个独特的自然光眼。柔和的光线透过屋顶和墙体结合处的天窗洒在司祭席和后面唱诗班处，吸引教众的目光。在这儿能看到的唯一外界景象就是天空，这样做是为了让人们专注于内心的沉思。

和唐博斯克教堂一样，位于捷克共和国的圣温塞斯拉斯教堂（124页）也是圆柱体，但这个圆柱体立面经过雕琢切割，某些地方可以

The tiny Votive Chapel in Como, Italy (p.14) is a pure piece of three-dimensional geometry, triangulated like a folded white napkin. Small, circular apertures in the vertical wall planes filter points of light, like celestial stars, representing the constellation Cassiopeia and the Polar Star. From certain angles the chapel seems to totter miraculously on a single surface edge. As in Msheireb Mosque and Agri Chapel, here again the logic of geometry and strict symmetry has been used.

Don Bosco Church (p.172) in Slovenia also forms part of a larger complex for the Salesian community. The complex has been planned around a courtyard with an old linden tree surrounded by tall delicate columns supporting a covered walkway. The layout of ecclesiastic spaces around a central courtyard follows the format of ancient European monasteries. The focus of the complex is the curved masonry drum, which is softened by the rhythm of lighter glazed bricks and adorned with a delicate concrete crown, reminiscent of Christ's crown of thorns. Within, the church hall is a concrete cylinder lit from above by a great circular light gun, a singular oculus of natural light. Skylights on the roof/wall junction wash muted light down into the presbytery and the rear choir, drawing the eye of congregants. The only view of the outside world is the sky, emphasizing inward contemplation.

Like Don Bosco, the St. Wenceslas Church (p.124) in the Czech Republic is also a cylinder, but here the cylinder has been sculpted, in places peeled away to allow indirect light to wash into the interior and to create an upper floor balcony. The circle was chosen. A site was carefully chosen where the church would be the focal point of the village,

斯科巴村民活动中心，斯洛文尼亚
Skorba Village Center, Slovenia

让间接的阳光洒进教堂内部，并在上层形成阳台。建筑师选择了圆形的建筑形式。该教堂的位置也是经过精心挑选的。在这里，教堂是整个村庄的焦点，与那些作为明显的地标建筑的老教堂相呼应。一个大大的十字架固定在屋檐的最高处，显示出该建筑作为教堂的功能。教堂内精致的木质靠背长凳的曲线与该建筑鼓形曲线一致，环绕着半圆形的后殿，突出了此处作为教会信徒集会场所的作用。屋顶上有一个三角形天窗，日光透过这个三角形天窗照射下来。

再如，斯洛文尼亚的斯科巴村，这个小村庄的居民决定在该村主干道的十字路口处建一个村民活动中心（24页），用来举办各种社会活动。整座建筑由白色混凝土建造，呈三角形，但三个角所处的位置高度各不相同。公共空间地面铺设了铺地材料。小礼拜堂的体量和看台被抬高，形成一个内向型村广场。该建筑大胆的外观足以彰显该地区的重要性。

上述建筑虽然方式各异，但都展现出建筑师如何利用光、空间和形式来创造出进行祈祷和冥想的空间。这些敬思空间的设计告诉我们应该如何利用所处位置的乡村景观环境来增强这些宗教建筑的力量，如何通过隐喻方式运用材料与形式、光和水，并赋予它们更大的意义。

echoing the historical position of churches as visible landmarks. A large cross is held aloft at the highest point of the eaves proclaiming the building's function. Delicate timber pews follow the curve of the drum embracing the curved apse and emphasizing the congregation as a communio of believers. Daylight filters down from above through a triangular skylight.
For another example, the residents of the small village of Skorba, Slovenia decided to erect a village center (p.24) for events and social activities, at the crossroads of the primary routes through the village. The entire structure is made of white concrete. In a triangular shape, its three corners are raised to different heights. The paved surface defines the communal space. The volumes of the chapel and the grandstands are raised to create an introverted village square. Its appearance is sufficiently bold to mark the significance of the area.

The above examples illustrate in ways how architects model light, space, and form to create spaces conducive to prayer and meditation. These places of worship show how the strength of buildings is often enhanced by the rural landscapes where they are located, how material and form, and light and water are used metaphorically and take on higher meanings.

1. Stegers, Rudolf, Entwurfsatlas, Sakralbau, Birkhäuser Verlag: Basel, 2008.
2. Stock, Wolfgang Jean, Architectural Guide, Christian Sacred Buildings in Europe since 1950, Prestel: Munich, 2004.

虔诚礼拜堂
The Votive Chapel

Mario Filippetto Architetto

一条新路的施工切断了一个富有历史性和宗教氛围的场所。虔诚礼拜堂建于18世纪末，是一座供奉圣母玛利亚的礼拜堂。如今，它被困在一条新的环形路的中央，人们根本无法进入。当地政府决定采取行动来解决这个问题。政府决定老礼拜堂仍留在环形路的中央，让其发挥历史纪念碑的作用，并另寻他处建造一座新的虔诚礼拜堂，给宗教团体提供一个可以继续礼拜并举行传统宗教仪式的场所。政府选择的新地点是附近的一座山，山上有一座纪念碑。这座山是当地社区的中心，风景秀丽，交通便捷。这些特点神奇地结合在一起吸引了政府的注意。新虔诚礼拜堂的基本设计概念很简单：维持老礼拜堂建筑结构的比例不变，但光线和体量的动态效果与老礼拜堂相反。新礼拜堂正面的最高点（金属十字架的顶点）和最低点（屋顶排水沟）的设计与老礼拜堂的相关尺寸相同，建筑楼面尺寸仍为3m×3m，但整座建筑旋转了45°以突出透景线。不同的点通过倾斜的切割方式连接起来，切割形成了顶部的天窗和正面的入口。老礼拜堂的纯白色也被保留了下来，以增强采光的效果并凸显建筑的神圣感。礼拜堂周围环境的设计在几何形状上与建筑本身的形状相匹配。环绕礼拜堂摆放的石头长凳，就像是它的边界。白色的砾石步行道直接通向白色大理石圣坛，修复后的圣母玛利亚油画像就供奉在圣坛后方。圣坛形状为平行六面体，上面有当地雕塑家雕刻的象征着天体和星体的符号。最后一处具有宗教象征意义的是圣坛两侧的小洞，这些小洞组合起来象征着仙后星座（代表圣母玛利亚）和北极星（代表上帝）。

The construction of a new road suddenly cut off a site of historical and religious preeminence: The Votive Chapel, built and dedicated to the Madonna at the end of the eighteenth century, was left stranded in the middle of a new roundabout and rendered completely inaccessible. The local administration took action to resolve the issue, deciding that the old Chapel would remain in the roundabout to serve as a historical monument, and pursuing the construction of a new Votive Chapel at another location to provide the religious community with a place to continue their worship and their established rituals. The new location they chose was a nearby hill occupied by a memorial monument. A center of the local community, the hill boasted beautiful landscapes and ease of access – a magical combination of features that captured the administration's attention. The basic concept for the new Votive Chapel was simple: the proportions of the old structure would be retained, but the dynamics of light and mass would be reversed. The new Chapel's tallest frontal point (over the metal cross) and

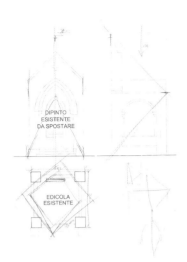

courtesy of the architect

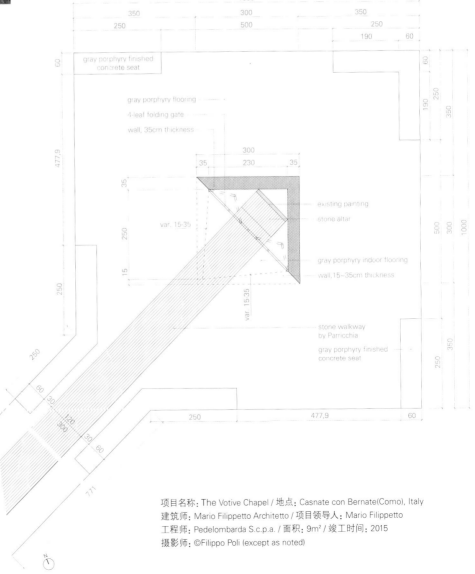

项目名称：The Votive Chapel / 地点：Casnate con Bernate(Como), Italy
建筑师：Mario Filippetto Architetto / 项目领导人：Mario Filippetto
工程师：Pedelombarda S.c.p.a. / 面积：9m² / 竣工时间：2015
摄影师：©Filippo Poli (except as noted)

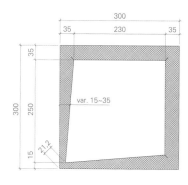

+3.00m平面图 +3.00 level floor

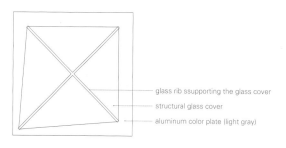

屋顶 roof

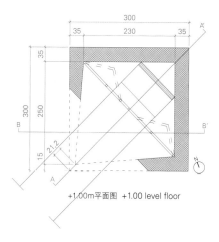

+1.00m平面图 +1.00 level floor

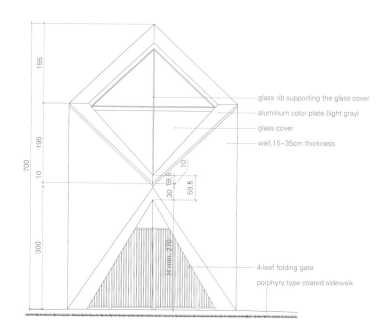

西立面 west elevation

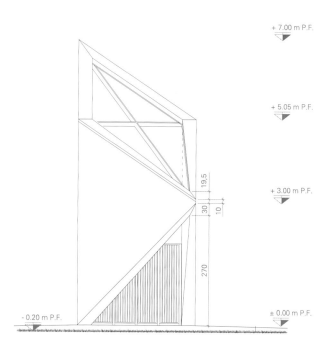

西北立面 north-west elevation

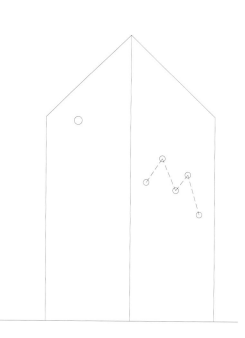

东立面 east elevation

lowest point (the roof gutter) have been designed with the same relative dimensions as the old Chapel; the 3-meter-by-3-meter floor area would also be preserved, but rotated by 45 degrees to emphasize the perspective lines. The points are connected by sloping cuts that form a skylight at the top and a front access opening. The old Chapel's pristine white color has also been preserved to emphasize the shapes of the light and the sacred nature of the building. The structure's surroundings are arranged geometrically in line with the building itself, with stone benches placed like a border around it. A white gravel walkway leads directly to the white marble altar, behind which is enshrined the restored painting of the Madonna. The altar is a parallelepiped with celestial and astral symbolism carved into it by a local sculptor. Finally, the religious symbolism is capped off with holes on the lateral sides representing the constellations Cassiopeia (representing the Madonna) and Polaris (representing God).

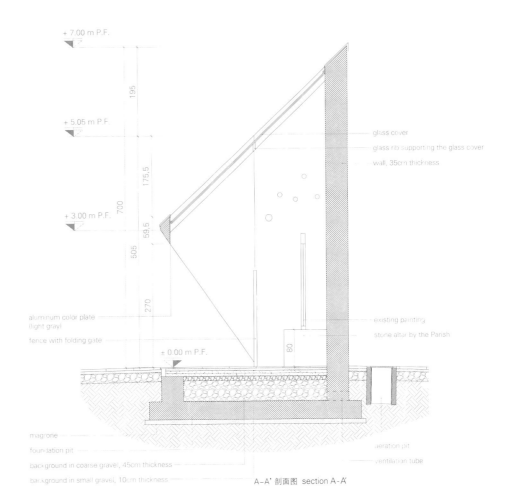

A-A' 剖面图 section A-A'

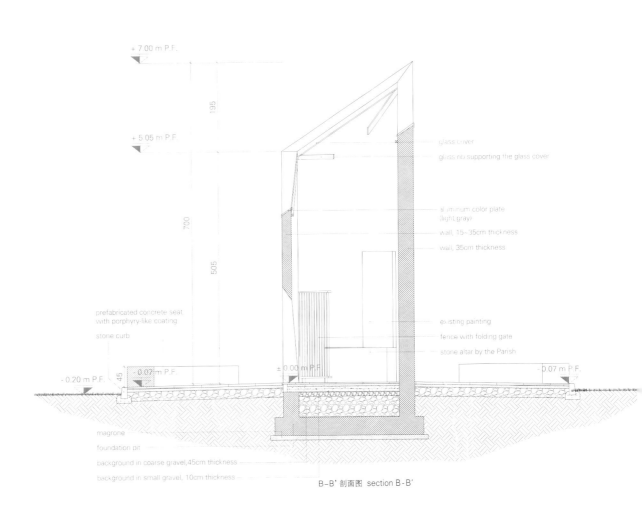

B-B' 剖面图 section B-B'

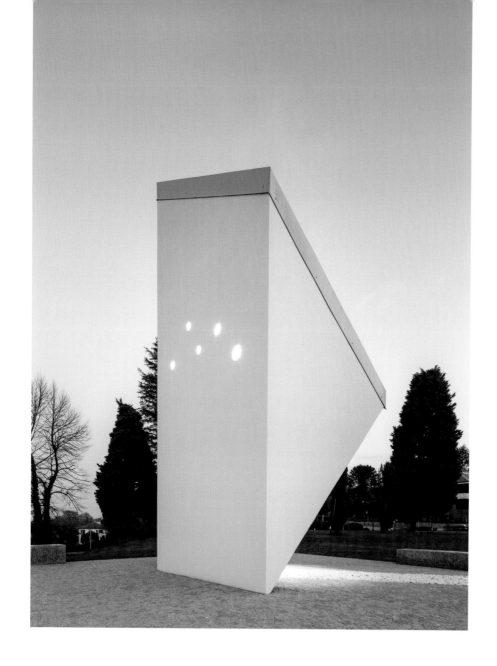
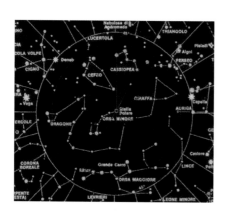
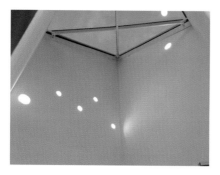
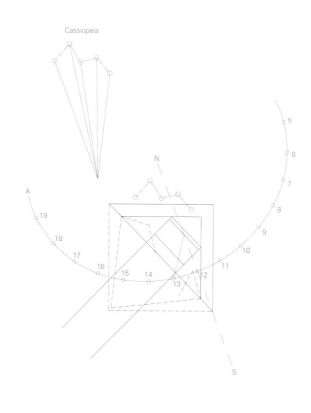

Cassiopeia

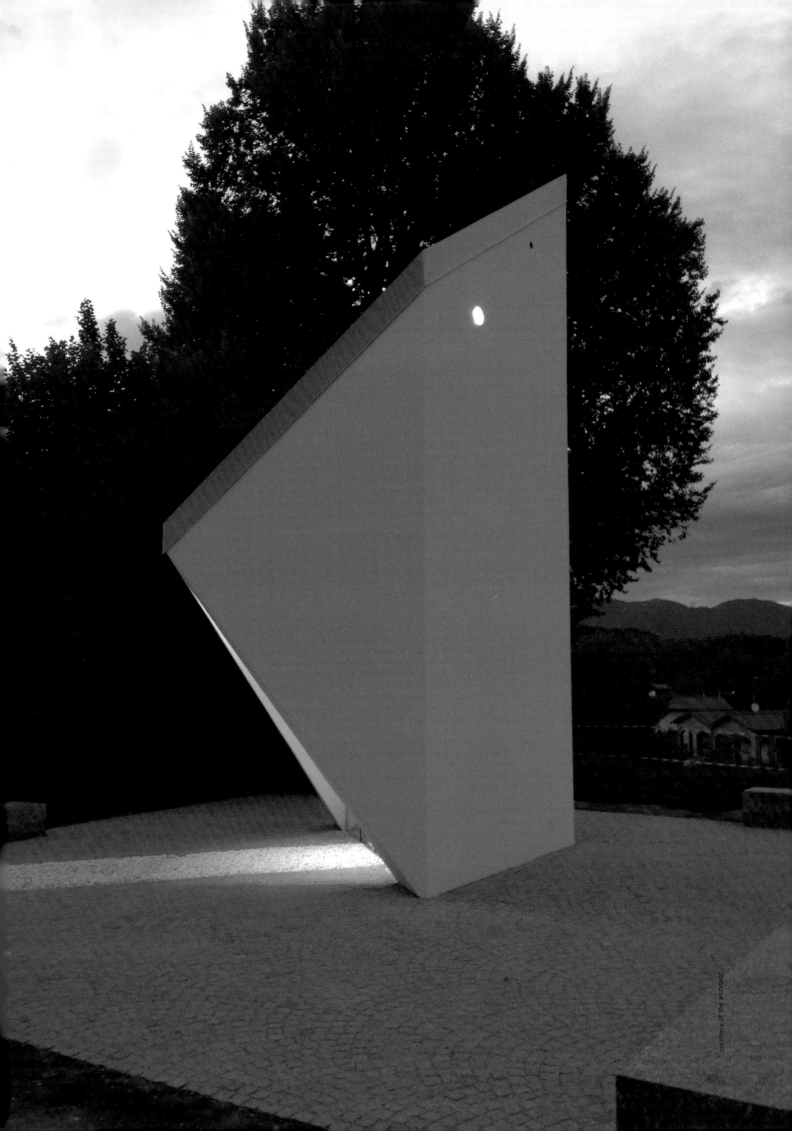

斯科巴村民活动中心
Skorba Village Center

ENOTA

斯科巴村位于斯洛文尼亚最古老的城镇普图伊附近。这里曾经是典型的聚居型村落,但是,随着岁月的流逝和城市化的发展,这种村落模式逐渐瓦解,成了一个没有清晰结构的用于通勤的郊区。协调的生长形式使村落沿着道路发生了显著的变化,形式混杂。这里没有公共区域,也没有功能清晰可辨的村民活动中心。

因此,在决定建造一座新的村庄礼拜堂之后,建筑师首先考虑的是如何利用这一机会为村民打造一个中央活动场所和社交空间。新村民活动中心的场址经过了精心挑选:这块场地曾经是村庄溪水的源头,也是河流旁的一块阶地开始上升的地方。新的村民活动中心就位于村庄主干道的十字路口,紧靠社区中心大楼。

在讨论斯科巴新村民活动中心项目时,值得注意的是,从初步的设计概念到最终的落成,整个项目都是村民自发推动完成的。他们积极地筹措资金用以购买地块,主动参与施工,长时间地无偿工作。在起草项目时,这些问题都必须经过仔细考量。

因此,新村民活动中心的设计并不依赖于每个细节的完美,也不依靠任何高科技手段博人眼球。相反,它的目的是主要通过清晰易懂的体量和简单明了的材料与周围环境建立适宜的关系。

界定空间的第一步是在空地中央的新广场中选定一个三角形的平面区域。铺砌的路面与草地明显区别开来,界定了未来的社交空间,并通过一条狭窄的步道与旁边的公路相连。两个铺砌面围合成中央略微下沉的活动空间。这种设计既可以使其免受周围环境的干扰,又可以使人们的目光聚集于此。接着,教堂和看台从地面抬升,形成一个内敛的村民广场。设计新村民广场的最后一步是使用统一的平面图确定抬升的体量,给人一种虚拟屋顶的感觉,完善了建筑的结构形式。

整个结构由相同的白色混凝土材料构成。简单的材料与鲜明的体量共同形成一种引人入胜的空间元素。它的外观足够大胆,足以使周围异质的环境逊色,显示着该地区的的重要性。

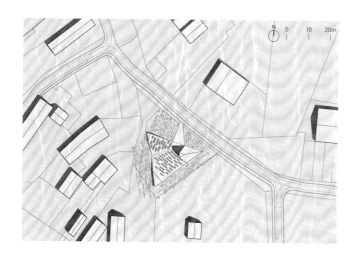

Skorba is a small village in the vicinity of Ptuj, Slovenia's oldest town. Once a typical village with a clustered settlement pattern, the passage of years and the proximity of the city caused it to grow out of turn, transforming it into a commuter suburb without a clear structure. The organic growth resulted in a markedly heterogeneous development organised along the access roads, with no public surface layout and without a clearly legible village center. Consequently, the first consideration following the decision to erect a village chapel was how to use this opportunity to also lay out a central event and socializing space for the residents. The site for the new village center was carefully chosen: the plot once contained the village stream source, and this is also the point where the one river terrace bank rises upwards to the other one. The village center is sited close to the community cental building, at the crossroads of the main paths through the village.

In discussing the new village center Skorba, it is remarkable

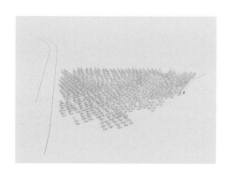
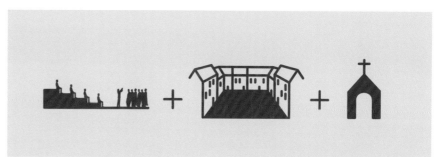
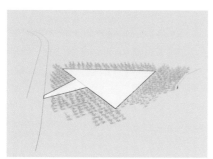
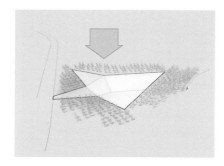
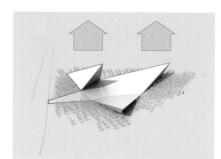
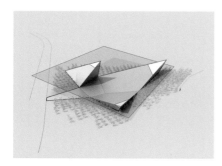
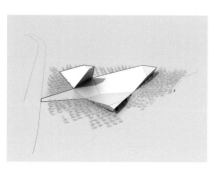

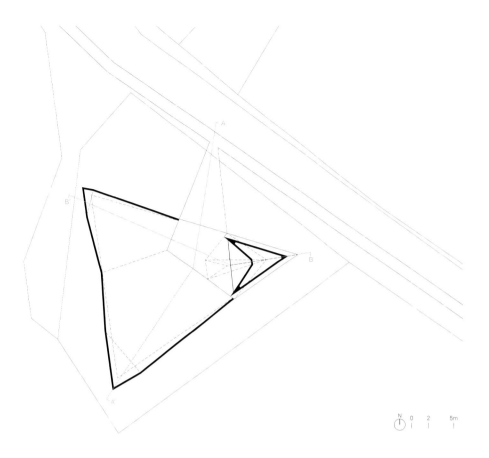

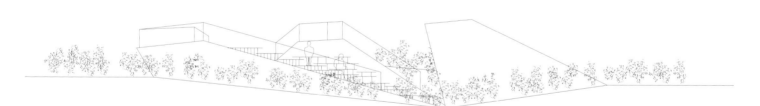

南立面 south elevation

东立面 east elevation

西立面 west elevation

that the entire project, from the initial idea to completion, was driven by the initiative of the inhabitants themselves. They took an active part in raising the funds to purchase the plot as well as in the construction, which required many hours of voluntary work by the villagers. All of this had to be taken into account when drafting the project.

Therefore, the design does not hinge upon the perfection of every detail and does not feature any high-tech solutions. Instead, it aims to establish an appropriate relationship with the surroundings chiefly by means of a clearly legible volumetric design and simple materiality.

The first step in defining the space is the siting of the triangular surface of the new square in the center of the vacant plot. The paved surface, clearly separate from the grassy surroundings, defines the future socializing space. The surface then employs a narrow access path to connect to the road passing by. The central part, created by the section of the geometries of both paved surfaces, is given a slight dip, which shelters the event space from the impacts of the surroundings and directs all users' gazes towards the center. Next, the volumes of the chapel and the grandstands are raised to create an introverted village square. The final device in designing the new village square is the truncation of the raised volumes by means of a uniform plane which creates an impression of a virtual roof and completes the structure's form.

The entire structure is made of a uniform material – white concrete. The combination of simple materiality and emphasised volumes creates an attractive spatial element, its appearance sufficiently bold to drown the heterogeneous structure of the surroundings and mark the significance of the area.

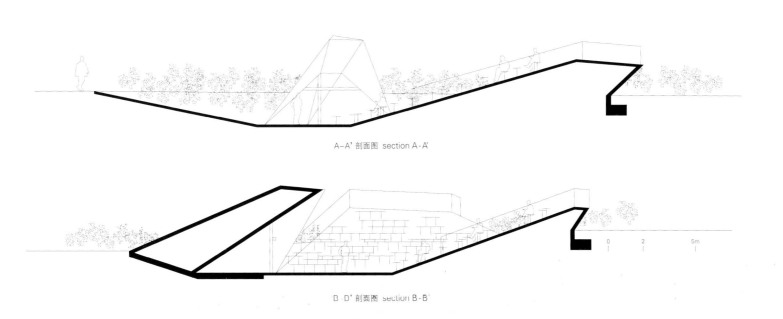

A-A' 剖面图 section A-A'

B-B' 剖面图 section B-B'

项目名称：Skorba Village Center / 地点：Skorba, Slovenia / 建筑师：ENOTA / 项目团队：Dean Lah, Milan Tomac, Nebojša Vertovšek, Alja Černe Mazalin, Nuša Završnik Šilec, Tjaž Bauer, Polona Ruparčič / 结构工程师：Elea iC / 总承包商（施工）：Aleksander Lončarič s.p. / 承包商：Mitja Simonič s.p. - mechanical installation; Franc Černelč - sculptor; Štefan Topolovec - artisan blacksmith; Tomaž Vratič s.p. - groundwork; Dušan Pernek s.p. - electrical installations; Miran Fekonja s.p. - concrete finishing / 供应商：Geoprojekt Zagreb, quarry in Sv. Ivan, Buzet (white sand for concrete, stone for sculpture) / 客户：Hajdina Municipality / 用地面积：1,430m² / 建筑面积：300m² / 总楼面积：300m² / 造价：120,000 EUR / 设计时间：2011 / 竣工时间：2017.9 / 摄影师：©Miran Kambič (courtesy of the architect)

法蒂玛圣母礼拜堂
Our Lady of Fatima Chapel

Plano Humano Arquitectos

　　法蒂玛圣母礼拜堂矗立在庄严的高原上,独享四周壮观的田园风光。该礼拜堂为了纪念法蒂玛圣母而建,其设计灵感来源于童子军的日常活动元素:户外野营、帐篷以及简约朴实的建筑风格和生活方式。尖锐的建筑边缘象征着童子军的领巾,也象征着童子军坚守誓言与承诺的精神。

　　礼拜堂的设计可以被看作是一个大型的经典帐篷,向所有人开放。其造型简约,山形墙屋顶欢迎着来自四方的来访者。建筑本身在较为低矮狭小的入口处离来访者最近,然后整个空间向前向上延伸,超出人们可以接触到的范围,人在穿行的过程中也能得到精神上的升华。礼拜堂四周令人眼花缭乱的景观放大了人们对礼拜堂的敬畏之感和惊叹之情。

　　礼拜堂为东西朝向,这使得日出的光辉能穿梭其中,也使得落日的余晖填满礼拜堂。各种色彩、色调和氛围交相辉映,与建筑共同呈现出一场视觉盛宴,令人赏心悦目。礼拜堂入口的设计形如童子军脖子上的领巾,在水道的映衬下更显独特。这条小小的沟渠贯穿整座建筑,召唤着来访者来此拜访和探秘。

　　水道穿越整座礼拜堂,直抵祭祀圣坛,然后融于无尽的景观之中。水道指引来访者走向礼拜堂外侧的十字架,而水道恰好将祭台和十字架连成一线。十字架突出了其背后的景观。它自身代表的神圣和自然景观融为一体,让来访者为之震撼。

　　礼拜堂内部由12根木梁(象征着耶稣的十二个门徒)支撑,裸眼可见,象征朴素求真的精神。

　　教堂总长度为12m,最大高度达9m。主梁在祭台上方升高,增加了室内空间的广度和深度,也进一步突出了这个神圣的地方。

　　材料的恰当选择使教堂与环境、童子军精神和建筑理念很好地融合在一起。童子军活动中经常会使用木材,所以人们看见木材就能自然地联想到童子军。这种传统而自然的材料具有坚硬和舒适的特点。锌材也是一种用于建筑的传统材料,它不但密封性强,还可以给人带来安全感。

不管是建筑层面还是宗教表达层面，光线都是一个重要的主题，因此教堂中的照明设计突出强调了建筑内部和外部的表现力。每当夜晚灯光亮起，伴随着天上的星星，礼拜堂就会熠熠生辉。

有一束光与众不同，它径直落在祭台上。

法蒂玛圣母礼拜堂为童子军团体服务，也可用于更大的团体举办庆祝活动。举办大型活动时，教会会众可以充分利用外部更大的景观空间，从而将礼拜堂本身变为一个圣台。

The Our Lady of Fatima Chapel stands privileged on a dignified plateau, boasting spectacular panoramic views of the rural environment. Dedicated to Our Lady of Fatima, the chapel is inspired by Scouting activities – from outdoorsmanship to camping, tents, and by humble, simple buildings and lifestyles. The building's pointed edges are a homage to the Scout's scarf, a symbol of the Scout's vow and commitment.

The building was conceptualized as a large classic tent whose doors are open to all. It is simple in design with a gable roof that welcomes all visitors. The structure itself reaches out to visitors at the low, narrow entrance, before stretching forward and upward beyond human reach and drawing visitors to something higher than themselves. The dazzling landscape in the background serves to amplify the awe and wonder of the chapel.

The chapel's east-west orientation means that the rising sun illuminates its interior and the setting sun fills the structure with a mélange of colors, tones, and ambiences that work in tandem with the architecture while pleasing the eye.

The scarf-inspired entrance, built to appear as though resting on a Scout's neck, is marked by the start of a watercourse. The small canal runs through the structure and serves as an invitation to the chapel and the Mystery that it celebrates.

项目名称：Our Lady of Fatima Chapel
地点：National Scouts Activities Camp, Idanha-a-Nova, Portugal
建筑师：Pedro Ferreira, Helena Vieira – Plano Humano Arquitectos
设计团队：Pedro Ferreira, Helena Vieira, João Martins
工程师：Emanuel Lopes – Tisem; Amilcar Rodrigues – IdeaWood
客户：Catholic Scout Association of Portugal – Corpo Nacional de Escutas; Board of directors, Father Luís Marinho
总承包商：IdeaWood
用途：religious / 面积：100m² / 竣工时间：2017
摄影师：©João Morgado (courtesy of the architect)

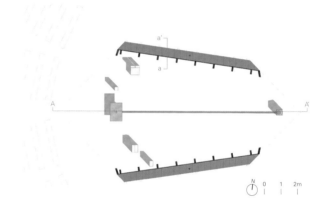

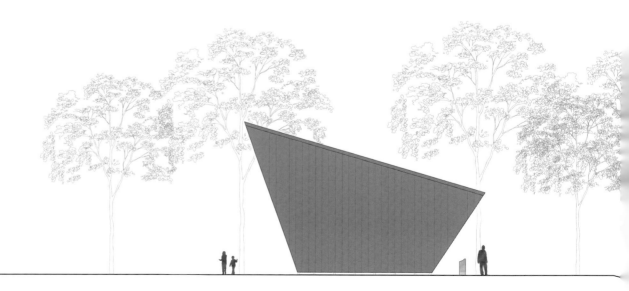

A-A' 剖面图 section A-A'

The watercourse runs through the entire length of the chapel, the structure marking its path transforming into an altar before joining the landscape. It directs visitors to the cross, which stands outside the chapel in alignment with the watercourse. The cross highlights the landscape behind it and strikes visitors with its scope and the way the scale of the landscape connects to the Divine.

The covering inside is supported by 12 wooden beams (an allusion to the Apostles), which are exposed to the eye to symbolize simplicity and truth.

Measuring at 12m in length, the structure reaches its highest point at the 9-meter mark, past the altar, at which point the main beam is raised to highlight the sacred area with the increased space and depth.

The building is steeped in its surroundings, the Scout community, and its architectural concept thanks to the materials that compose it. Wood is naturally connected to the Scouts, being used often in their activities. The fact that it is a traditional and natural material provides a sense of solidarity and comfort. Zinc is another traditional material that composes the structure, chosen not only for its tightness but also for the feeling of protection it offers.

As light is an important theme in both architecture and religious expression, the lighting in the structure was designed to highlight the expressive nature of both the interior and the exterior. At night, the building is illuminated as the light frames the structure with the stars overhead.

One beam of light was designed to stand out from the rest and fall directly on the altar.

Our Lady of Fatima Chapel serves the Scout community and hosts celebratory events for larger groups as well, in which case the congregation may make the best of the larger exterior space adjoining the landscape, turning the chapel building itself into an altar.

- VMzinc plate coating, in anthra color
- OSB board of 18mm thickness
- glued laminated pine wood beam
- metal knee-cap
- concrete foundation

- VMzinc plate coating, in anthra color
- studded screen insulation
- OSB board of 18mm thickness
- glued laminated pine wood beam
- metal knee-cap
- glued laminated pine wood beam
- gravel
- existing ground
- concrete foundation
- wide grading Tout-Venant

a-a' 详图 detail a-a'

水岸佛堂位于中国河北省唐山市，它的周围有绿水茂林环绕。这是一个供人参佛、静思、冥想的场所，同时也可以满足个人的生活起居需求。

这里沿河有一块土丘，背后是广阔的田野和零星的蔬菜大棚。设计从建筑与自然的联系入手，采用覆土的方式让建筑隐于土丘之下，同时又以流动的内部空间彰显自然的神圣气质，从而塑造出一个树、水、佛、人共存的，具有感受力的场所。

为了将河畔树木完好地保留下来，该设计小心地避开了所有树植，整座建筑的平面布局也像分叉的树枝一样在原有的树林之下延伸。南北走向与沿河方向的两条轴线将整座建筑划分为五个独立但又连续的空间。五个"分叉"代表了五种不同的空间：入口、参佛室、茶室、起居室、卫浴间，共同构成了一种漫步式的体验。建筑始终与大树和自然景观有着紧密联系。入口正对着两棵树。人从树下经由一条狭窄的小径缓缓走入佛堂。佛龛背墙面水，天光与树影通过天窗沿着弧形墙面柔和地洒入室内，彰显着佛祖的光辉。

茶室面朝着遍植荷花的水面完全敞开，几棵树分居左右，成为庭院的一部分，给人品茶与观景的乐趣。休息室与建筑的其他部分由一个竹院分隔，使得日常活动能随着一天中时间的变化而不同。覆土建筑整体成为土地延伸出来的一部分，是树荫下又一座可以供人们使用的"土丘"。

与自然的关系也可以进一步延伸到材料层面。建筑的墙体和屋顶采用混凝土整体浇筑，一次成型。混凝土模板由3cm宽的松木条拼合为一体。自然的木纹与竖向的线性肌理被映刻在室内墙面，给冰冷的混凝土材料增添了柔和温暖的感觉。嵌入式家具也是定制的，用木条制作，其灰色的木质纹理与混凝土墙略有不同。室内地面采用光滑的水磨石材料，表面有细细的石子纹路，将外界的自然景色映射进室内。

水岸佛堂

Waterside Buddhist Shrine

Archstudio

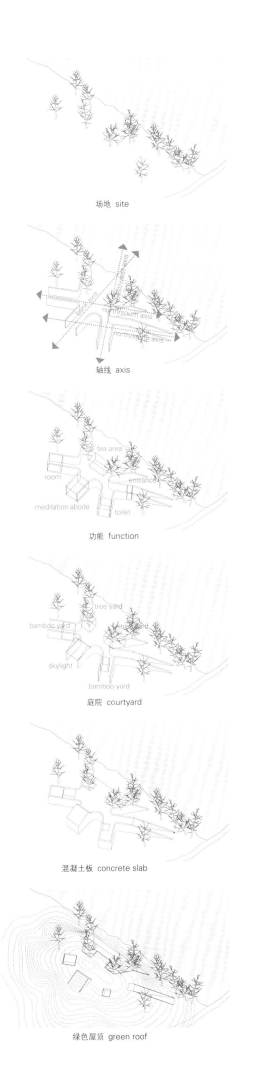

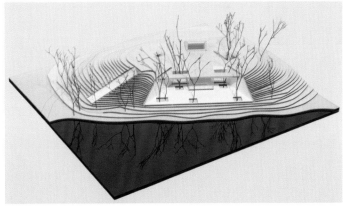
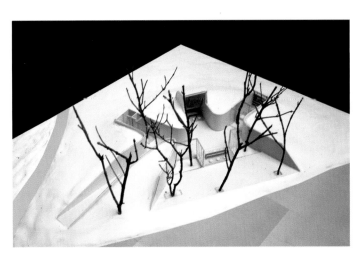

项目名称：Waterside Buddist Shrine / 地点：Tangshan, Hebei, China
建筑师：Archstudio / 设计团队：Han Wenqiang, Jiang Zhao, Li Xiaoming
结构设计：Zhang Fuhua / 水电设计：Zheng Baowei
用途：religious building / 用地面积：about 500m² / 建筑面积：169m²
设计时间：2015.4 – 8 / 施工时间：2015.10 – 2017.1
摄影师：©Wang Ning (courtesy of the architect) - p.46~47, p.50~51, p.52, p.53, p.54, p.56~57; ©Jin Weiqi (courtesy of the architect) - p.44, p.49

室外地板采用白色鹅卵石与水泥浆砌成，因此室内地板与室外地板的触感不同。所有门窗均使用了实木材料，以体现材料的自然质感。禅宗讲究顺应自然，融入自然。这同样也是这个空间设计的目标。它利用空间、结构、材料来激发人的感知，从使而人与建筑都能在平凡的乡村景观中发现自然的魅力，并与自然共生。

This is a place for Buddhist meditation, thinking and contemplation, as well as a place satisfying the needs of daily life. The building is located in the forest by the riverside of Tangshan, northeastern Hebei Province, China.
Along the river, here is a mound, behind which is a great stretch of open field and sporadic vegetable greenhouses. The design started from the connection between the building and nature, adopts the method of earthing to hide the building under the earth mound while presenting the divine temperament of nature with flowing interior space. A place with power of perception where trees, water, Buddha and human coexist is thus created.

To remain trees along the river perfectly intact, the building plan avoids all trunks. Shape of the plan looks like branches extending under the existing forest. Five separated and continuous spaces are created within the building by two axes, among which one is north-south going and another one goes along the river. The five "branches" represent five spaces of different functions: entrance, Buddhist meditation room, tea room, living room, and bathroom, which form a strolling-style experience together. The building remains close to trees and natural scenery. The entrance faces two trees; people need to walk into the building through a narrow path under the trees. The shrine is against the wall

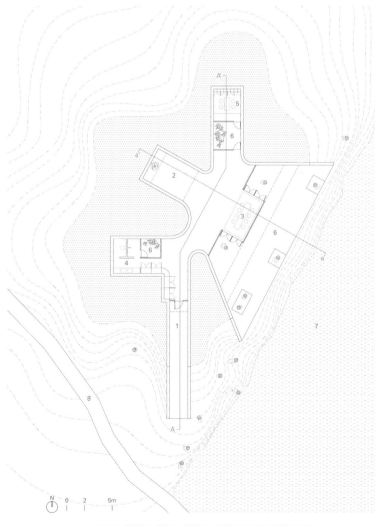

1. 入口 2. 参佛室 3. 茶室 4. 卫浴间 5. 起居室 6. 庭院 7. 河流 8. 道路
1. entrance 2. meditation room 3. tea area 4. bathroom 5. living room 6. courtyard 7. river 8. road
一层 ground floor

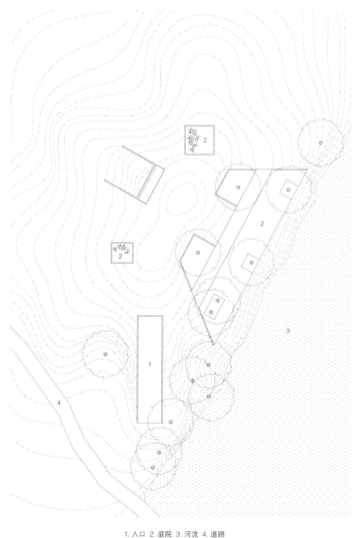

1. 入口 2. 庭院 3. 河流 4. 道路
1. entrance 2. courtyard 3. river 4. road
屋顶 roof

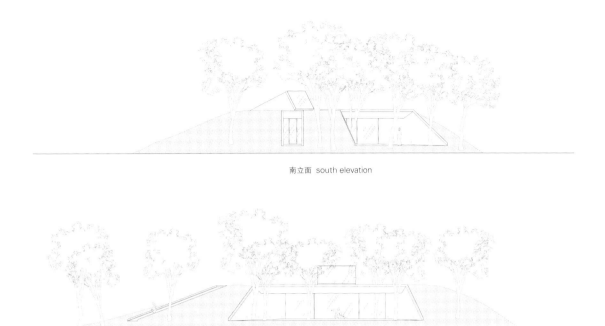

南立面 south elevation

东南立面 south-east elevation

and facing the water, where the light and the shadow of the trees get through the skylight and flow into the interior space softly along the curved wall, exaggerating the light of Buddha.

The tea room opens completely to the pool which is filled with lotus, and trees on both sides of the tea room has become part of the courtyard, creating a fun of tea tasting and sight-viewing. The lounge is separated from other parts of the building by a bamboo courtyard; such division enables daily life varies with different hours of a day. The whole building is covered with earth and becomes an extension of the land, as another "mound" which could be used under the trees.

The relationship with nature further extends to the use of materials. Integral concreting is used in walls and the roof of the building. The concrete formwork is pieced together with pine strips of 3cm width, in this way natural wood grain and vertical linear texture are impressed on the interior surface, creating a soft and warm feeling to the cold concrete materials. Built-in-furniture is custom-made with wood strips, whose gray wood grain is a little bit different from the concrete walls. Smooth terrazzo is used for the interior floor, where there is thin grain of stone on the surface, and it maps the outdoor natural landscape into the interior space. Cement grouting with white pebbles is adopted in outdoor flooring, which creates a difference in sense of touch between indoor and outdoor floor. To reflect natural texture of the materials, solid wood is used for all doors and windows. Zen stresses on complying with nature and being part of nature. That is also the goal of the design for this space-taking use of space, structure and material to stimulate human perception, thus helping man and building to find the charm of nature even in an ordinary rural landscape, and to coexist with nature.

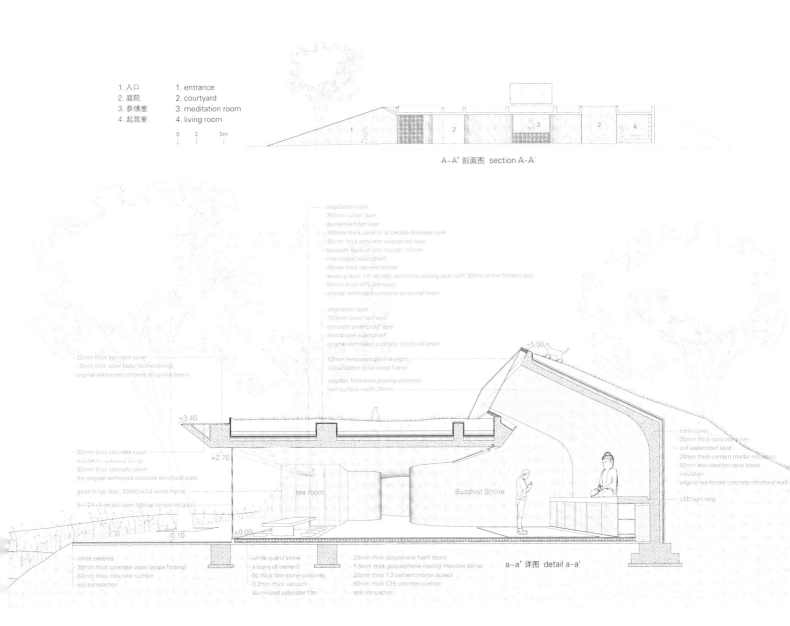

A-A' 剖面图 section A-A'

a-a' 详图 detail a-a'

Wirmboden 高山礼拜堂
Wirmboden Alpine Chapel

Innauer-Matt Architekten

在奥地利西部被称为布雷根泽瓦尔德的一处山谷处,牲畜季节性迁移放牧对牧民来说司空见惯:在温暖的季节里,牧民们会把牲畜赶到山间牧场去吃草。这里海拔较低一些的牧场被当地人称为Vorsäß,而海拔稍高一些的牧场被称为Alpe。海拔稍高一些的牧场常常在最温暖的夏季月份里用于放牧。

Wirmboden就是一处Vorsäß。Wirmboden牧场位于Kanisfluh山山谷陡峭的北坡脚下,由农民们集体拥有和管理。几乎每一处Vorsaäß都有自己的小礼拜堂,或者至少是有某个特定的空间,用于举行弥撒,或者用于举行农民和他们的牲畜传统的献祭仪式。

32年来,Wirmboden牧场一直有一个小礼拜堂,但在2012年,礼拜堂和几间小屋被雪崩摧毁。虽然从一开始就很清楚这些棚屋将被重建,但要就新礼拜堂的建设达成共识却极具挑战性。由于客户群体是一批农民,每个人对礼拜堂的重建都有自己的看法,所以困难并不在于建筑设计,而在于调和人际关系。起初,要找到一个令所有人满意的设计方案似乎是不可能的。

所以,我们现在在Wirmboden所看到的这座建筑,正是这个非常多样化的群体集体精神的象征。他们历时三年的协调、构思、规划,最终建造了这座礼拜堂——相当于每平方米建筑面积耗时6个月才完成。现如今,这座山上小礼拜堂的落成,让这个阿尔卑斯山的古村落更加完整,与之更是相得益彰,成为邻居们平时聚会、集会和举行庆祝活动的地方,成为人们祈祷片刻的地方。

这栋新的礼拜堂遵循极简的设计原则,呼应了最原始的宗教建筑形态,突出了建筑的特殊身份和用途。按照传统,墙体是由从周围收集的石头和夯实的混凝土建成的,陡峭的桁架上覆盖着粗糙的木瓦片。小礼拜堂的入口处是用木材建造的,非常狭窄,引导人们进入里面小的祈祷室祈祷。

礼拜堂内部简单朴素,它最重要的作用就是用来举行仪式庆典和为人们提供一个自我反思的空间。漫射的阳光透过屋脊上的一个开口洒进室内。这个天窗开口由喷砂玻璃和喷砂不锈钢制成,与浅蓝色的祭坛窗户相映成趣,营造出一种空灵飘渺、冥想沉思的氛围,用于纪念

的照片卡放置在椽子之间的狭窄空间处，用于纪念Wirmboden人民挚爱的人。入口、桁架和入口上方带有钟铃的空间都是由德国云杉（有时也被称为榛子云杉）建成的。这种木材因其特殊的声学特性通常被用于制作小提琴和吉他。

正是由于Wirmboden每个成员的努力和贡献，新礼拜堂的建造没有任何第三方的帮助。全民参与使一开始看似不可能的事情成为现实，结果几乎让所有人都满意，这座礼拜堂也成为山中瑰宝。

In the Bregenzerwald, an alpine valley in Western Austria, transhumance is still the commonly practised form of farming: Livestock is driven to graze on mountain pastures in the warmer months. The lower ones of these pastures are called Vorsäß while the higher ones, used in the warmest summer months, are called Alpe.

Wirmboden is a Vorsäß at the foot of the steep north face of the valley's Kanisfluh mountain, owned and managed by a collective of farmers. Almost every Vorsäß has its own little chapel or at least some designated space for the celebration of masses and the traditional consecration of the farmers and their livestock.

For 32 years there was a little chapel at Wirmboden, but in 2012, the chapel and several huts were destroyed by an avalanche. While it was clear from the beginning that the huts would be rebuilt, it was more challenging to find a consensus on the construction of a new chapel. With the client being a collective of farmers, each with their own, differing opinion, the difficulties were not so much of architectural but rather of interpersonal nature. Finding a solution

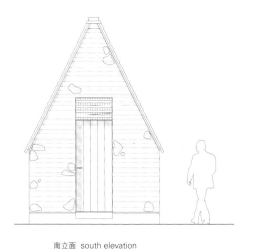

南立面 south elevation

东立面 east elevation

北立面 north elevation

that would make everyone happy seemed impossible in the beginning.

So what we see now at Wirmboden is a symbol for the collective spirit of this very diverse group of people. It was them who negotiated, conceived, planned, and eventually built this chapel over the course of three years – that's 6 months for every square meter of floor space in the building. Today, the mountain chapel complements the ensemble of alpine huts most naturally; it became the place where neighbours meet casually, where gatherings and celebrations are held, where people come to take a moment and pray.

The simple, basic outline of the new chapel refers to the most original form of sacred buildings and highlights the characteristics of this special place and its use. According to tradition, the walls are made from stones collected around the place and tamped concrete. Rough split shingles cover the steep truss, and a narrow wooden entry leads into the small oratory.

With its simple and humble interior, the chapel is first and foremost a place of commemoration and reflection. Diffuse daylight falls through an opening in the roof ridge in blasted stainless steel to play with the light blue altar window, creating an ethereal, contemplative atmosphere. Memorial photo cards are placed in the narrow spaces between rafters, commemorating loved ones of the Wirmboden people. Entrance, truss and the bell space above the entrance are made from German spruce (sometimes called hazel spruce), a type of wood that is normally used for violins and guitars for its special acoustic qualities.

Thanks to practical contributions by almost every single member of the Wirmboden collective, the new chapel was built without any help from third parties. Everyone's participation made possible what seemed impossible in the beginning – making (almost) everyone happy with the result: A gem in the mountains.

sandblasted glass
roof ridge, girder:
stainless steel sandblasted/polished

rough split shingles
spruce battens
rafters German spruce

temped concrete/collected stones
casting made from rough spruce planks

concrete/collected stones
smooth finish

0 20 50cm

详图1 detail 1

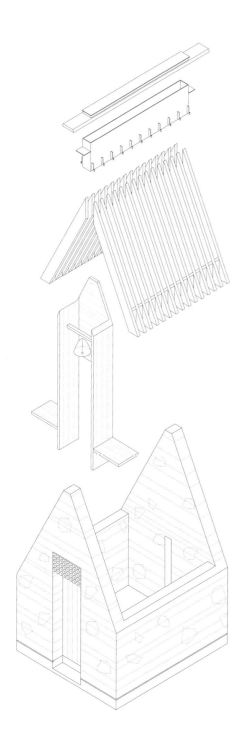

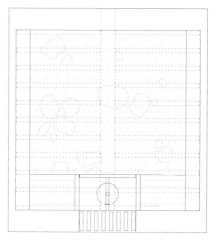

屋顶 roof

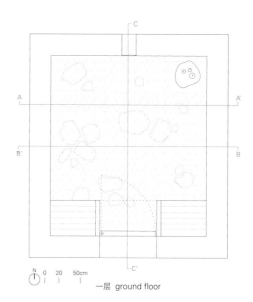

一层 ground floor

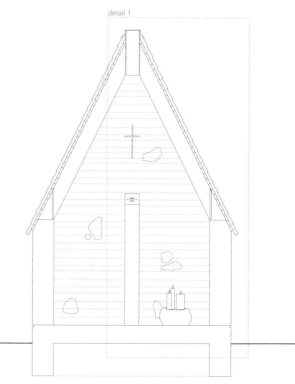

A-A' 剖面图 section A-A'

项目名称：Wirmboden Alpine Chapel
地点：6882 Schnepfau, Vorarlberg, Austria
建筑师：Innauer-Matt Architekten
静力学设计：Gordian Kley-Merz Kley Partner
施工监理：Karlheinz Gasser
客户：Wirmboden collective of transhumance farmers
建筑面积：6m²
结构：solid construction from collected stones and tamped concrete, truss with narrow spruce rafters fit for heavy snow load, ridges in blasted stainless steal; single glazing, blasted, roofing in rough split shingles, single-glazed altar window in light blue
室内饰面：benches, bell frame, entrance door in solid German spruce, fixtures, keys, handles, cross in blasted stainless steel
地板饰面：hand-smoothed concrete with collected flat stones
开发时间：2013—2016
施工时间：2016.4—8
摄影师：©Adolf Bereuter (courtesy of the architect) (except as noted)

©Kairheinz Gasser (courtesy of the architect)

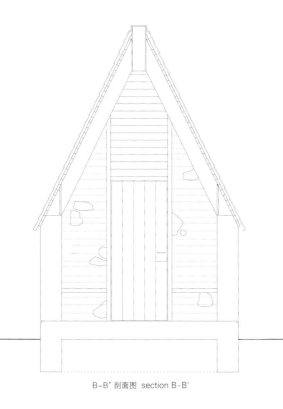

B-B' 剖面图 section B-B'

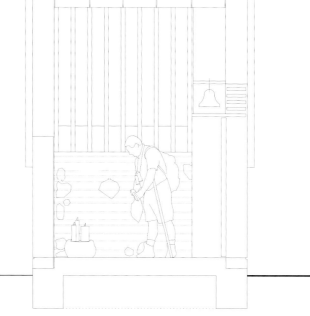

C-C' 剖面图 section C-C'

68

圣伊利教堂
Saint Elie Church

Maroun Lahoud Architecte

圣伊利教堂矗立在Brih市的Mtayleh,这儿属于黎巴嫩山地区的舒夫区。Brih市离黎巴嫩首都贝鲁特约53km。这儿地势属于梯田形坡地,有着丰富的植被。该地区还有一段复杂的历史,1975年至1990年内战期间,暴力冲突不断,导致这里的房屋和礼拜场所几乎完全被摧毁,村民们流离失所。圣伊利教堂是该地区第一个象征冲突双方和解的项目。

该项目由教堂建筑和周围的广场组成,旨在把人们聚集在一起来感恩自然。圣伊利教堂通体被装饰成白色,坐落在由石墙围挡的、看似枯燥乏味的基座之上,显得庄严而肃穆。教堂改变了山坡的地形,它由多功能大厅和一些附属建筑组成。

教堂立面由经过凿石锤打磨的白色石材装饰,显得熠熠生辉,象征着重生。纯粹的体量和平屋顶等方面的设计都体现了马龙派教堂的特征。圣伊利教堂的方形底座为17m×17m,一次最多可容纳250人。圣器室和忏悔室位于教堂后部,使圣坛周围的空间最大化。

建筑外墙上的石头长度从25cm到45cm不等,其排列虽没有固定的模式和图案,但给人整齐划一之感。这种统一感也是钟楼和下面教堂入口的设计理念;厚重的入口是一个象征性的过滤器,将神圣世界与外面的世俗世界区分开来。

自入口处,来访者的目光就不由自主地被吸引到一个颇为宏伟的十字形狭缝窗户上;因为它面向北方,所以避免了弥撒过程中任何不必要的背光效果。在这座充满自然光的盒子式建筑里,白色的墙壁似乎漫射着自然光线,而乳白色、有机的卡拉拉大理石地板反过来又可以反射这些自然光。

室内设计旨在通过间接照明方式来定义其神圣维度,从而散发出一种精神的提升感;顶部照明有的设在圣坛、圣器室和忏悔室的上方,有的沿着墙体设计,还有的被小心翼翼地设在墙体背面。

该设计中窗户的数量具有宗教上的象征意义。三个光龛代表的是三位一体,总计有12个窗户,象征着耶稣基督的赴难之路。

圣伊利教堂象征着当地社区历史上新的篇章,使用当地石材的做法使其深深根植于其渊源。

当地居民利用被拆除的房屋和教堂中废弃的材料,来协助建造基座部分。当地的工匠们也非常乐于去选择使用这些废弃的材料。

500m²的多功能大厅有五个大窗户,向庭院和周围景观敞开,为多功能厅提供自然采光。另外,几个设在建筑物空隙中的附属设施为多功能厅提供服务。建筑所在的位置和所使用材料的鲜明差异性,使该项目已成为舒夫区黄金谷一个新的焦点,也因为其尊重历史遗产而在黎巴嫩山地区的历史上写下了新的一页。

The Saint Elie church stands in Mtayleh, Brih in the Shouf District in Mount Lebanon. Brih, located 53 kilometres away from Lebanon's capital of Beirut, is characterized by its terraced topography and abundance of vegetation. The region is also marked by a complicated history, which led to violent clashes during the civil war between 1975 and 1990 that resulted in the near-total destruction of houses and places of worship, and the displacement of villagers. The Saint Elie Church is the first project symbolizing reconciliation in the region.

The project was conceived to bring people together in celebration of nature. It is comprised of the church building and the surrounding plaza. Dressed entirely in white and set solemnly on a dry stone walled base, the Saint Elie Church has transformed the topography of the hillside to house its multipurpose hall and annexes.

Radiant with its white bush-hammered stone cladding, the church inspires renewal. Aspects such as the pure massing and flat roof embody characteristics of the Maronite Church. With a square base of 17x17m, Saint Elie Church can house up to 250 people at a time. The sacristy and the confessional are located at the back to maximize the space around the altar.

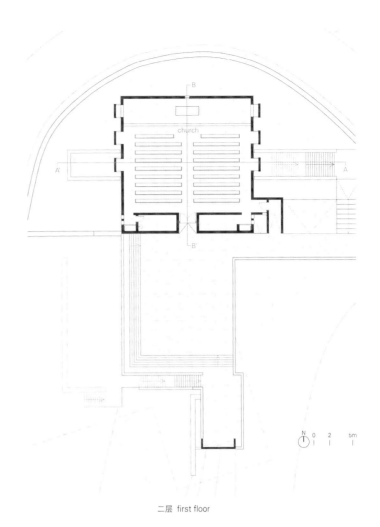

二层 first floor

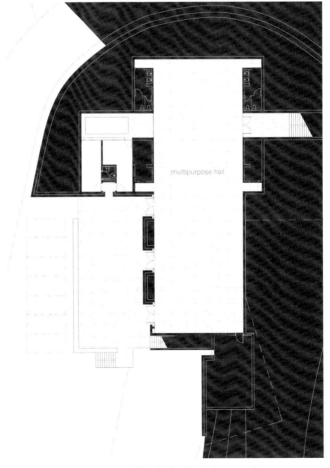

地下一层 first floor below ground

项目名称：Saint Elie Church
地点：Brih, Shouf, Lebanon (50km from Beirut)
建筑师：Maroun Lahoud Architecte
结构、技术顾问：Bureau International de Genie
合作者：Salam Geha, Dany Ajouz
客户：Ministry of displaced
建筑面积：950m² / 造价：1.4M USD
材料：local white bush-hammered stone, onsite stones, marble
设计开始时间：2014 / 竣工时间：2016
摄影师：Courtesy of the architect

The stones composing the exterior range from 25 to 45cm in length and have been lined without pattern while preserving a sense of unity. The same idea of unity was the driving concept for the bell tower and the church entrance built below; the thick entrance is a symbolic filter that sets the sacred world apart from the material world.

From the entrance, visitors are drawn to an imposing cross-shaped slit; because it is oriented towards the north, it avoids any unwanted backlight effect during the mass. In this natural light box, the white walls seem to diffuse light, with the milky, and organic Carrara marble flooring reflecting it in turn.

The interior is designed to exude a sense of spiritual elevation with indirect lighting schemes that define its sacred dimension: zenithal lighting has been positioned above the altar, sacristy and confessional, along the walls, and discreetly at the back.

The number of openings in the design symbolize liturgical doctrines. Three light niches along the lateral circulations refer to the Trinity; disposed on both sides, with two openings each, they amount to 12 openings in total, as a reference to the way of the cross.

The Saint Elie Church symbolizes a new page in the history of the community, but the use of stones from the local area keeps it rooted in its origins.

Locals assisted in the construction of the base by bringing the abandoned materials from their demolished houses and churches, and the local artisans readily made efforts to choose what to use from them.

The 500m² multipurpose hall is naturally lit by five large bays open into the courtyards and the landscape, and is served by several annexes gliding in the interstices of the structure. Due to its location and the contrast of its materials, the project has become a new point of interest in the Shouf's valley of Gold and writes a new page in the history of Mount Lebanon by honoring its heritage.

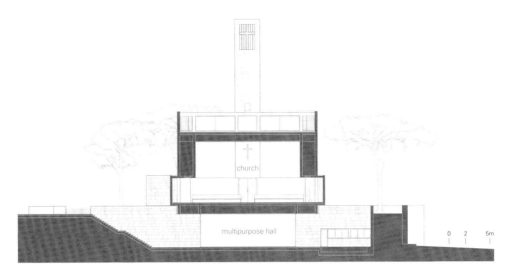

A-A' 剖面图 section A-A'

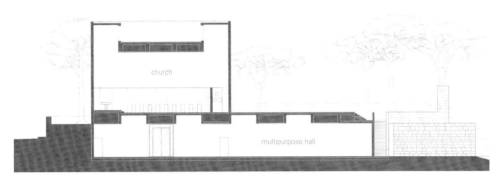

B-B' 剖面图 section B-B'

平衡礼拜堂
Chapel of Equilibrium

Álvaro Siza + Carlos Castanheira

庆尚北道的军威是一个小山村，位于首尔东南约250km处，南面可以看到八公山。Bugye植物园坐落在一个有2000人居住的区域的边缘。植物园内树木种类繁多，有温柏、银杏和松树等，其中一些树木已有200多年的历史。

这里还有一个凉亭、一个艺术展览馆、一个气象观测台以及一个供人小憩、静静欣赏周围景色的冥想空间。在植物园的北端，还有一座小礼拜堂，由葡萄牙建筑师阿尔瓦罗·西扎和他的长期合作伙伴卡洛斯·卡斯塔涅拉设计。礼拜堂是纯白色的，映衬着深绿的山峦，形成鲜明的对比。阿尔瓦罗·西扎的设计灵感主要来自当地的特点和周边环境，他会用心感受每一寸土地，将当地独特的价值和特点融入其设计之中，随之改变建筑风格。西扎曾经断言："建筑师不发明任何东西，他们只是改变现实。"

礼拜堂沿着倾斜的山脊向东延伸。清晨，第一缕阳光透过礼拜堂东墙上的一扇小窗洒满整个空间。该建筑由一个约2m高的小人字形顶棚，一个较高的人字形屋顶下方的过渡空间和一个较大的立方体空间组成。来访者沿着不断升高的天花板往里走，可以感受到一种富有戏剧性的空间变化。

来访者走进入口，眼前是突如其来的黑暗，他们会驻足停留。如果继续往前走几步，光明就会倾泻在头上，赶走黑暗。清除世间烦扰和心灵债务。仿佛一脚踏进这个礼拜堂，便开始了一种神圣的仪式。"小礼拜堂是一个充满感恩和奇迹的地方，是一个寻找神圣与宁静的地方，"卡洛斯·卡斯塔涅拉说。

日日变换，季季更迭，礼拜堂的白色也在大自然中发生着变化。盛夏，它映射着树叶的翠绿；深秋，它被涂抹上秋天的红晕；冬日，它似乎隐藏在皑皑白雪里。在雾蒙蒙的早晨，雾纱遮盖住真容；在夕阳西下时，余晖粉饰着它的脸颊。

恰到好处的白墙，陡斜的屋顶，使小礼拜堂与植物园融为一体，与风景融为一体。

Gunwi, Gyeongsangbuk-do, about 250km southeast of Seoul, is a mountain village where Mt. Palgong is visible to the south. Bugye Arboretum is located on the side of a small neighborhood of 2,000 people. The Arboretum boasts a variety of tree species such as quince, ginkgo and pine, some of which are over two hundred years old. There is a gazebo, an art pavilion, an observatory, and a meditation space where one can sit for a while to relax and take in the surroundings quietly in tranquility. There is also a small chapel designed by Portuguese architect Álvaro Siza and his long-time partner Carlos Castanheira at the northern end of the arboretum. It is pure white, striking against the dark green of the mountain. Álvaro Siza is an architect

项目名称：Chapel of Equilibrium / 地点：Gunwi, Gyeongsangbuk-do, Korea / 建筑师：Álvaro Siza + Carlos Castanheira / 葡萄牙事务所：CC&CB Arquitectos
合作者：Rita Ferreira, Diana Vasconcelos / 顾问：HDP - Paulo Fidalgo (structure) / 3D模型：Germano Vieira / 总楼面面积：32m² / 设计时间：2015.11—2017.6
施工时间：2017.8—2017.12 / 竣工时间：2018 / 摄影师：©JongOh Kim

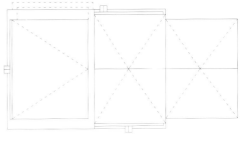

屋顶 roof

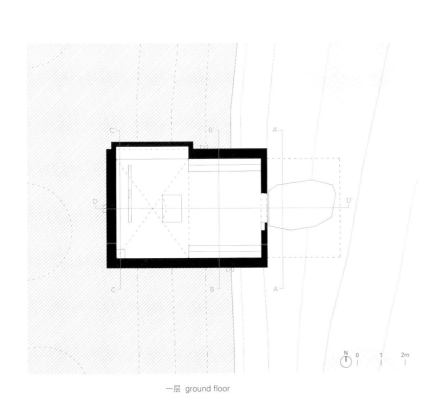

一层 ground floor

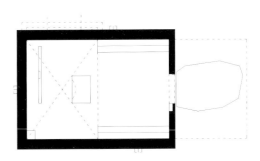

+6.50m平面图 +6.50m floor

who designs with local characteristics and surrounding environment. He reads the land and transforms the architectural style in accordance with the unique values and characteristics of the land. Siza once asserts that "architects don't invent anything, they just transform reality."

The chapel is oriented eastward along a sloping ridge. In the morning, the first rays of light slowly fill the space through a small window on the eastern wall. The building is made up of a small gable roof canopy about 2 meters high, a transitional space beneath a higher gable roof, and a larger cubic space. Walking along the ever-rising ceiling, visitors feel a dramatic sense of space.

Visitors walk into the entrance and pause in sudden darkness. When you take a few more steps, the light pours down over your head, clears the darkness and cleanses the burden of the world and the debt of the heart. It seems that just entering the building begins a godly ritual. "The chapel is the place to find the Divine and Peace with full of Gratitude and Wonder," says Carlos Castanheira.

As the time of the day and the season of the year change, the white of the chapel also changes in nature. It reflects the green of the leaves in high summer, the ruddy glow of autumn, and appears to be camouflaged in the snow. It hides its appearance behind a scrim of fog on misty mornings, and paints its face in a golden sunset when the sun goes down.

The modest volume of white walls and sharply sloping roofs melds with the arboretum itself, and soon became one with the scenery. Hyun Yu-mi

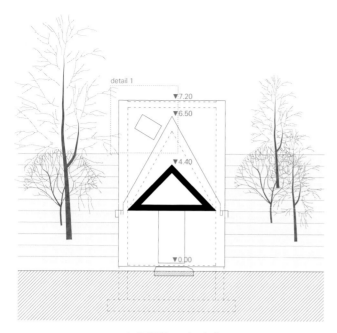

A-A' 剖面图 section A-A'

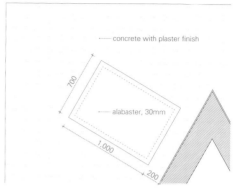

详图1 detail 1

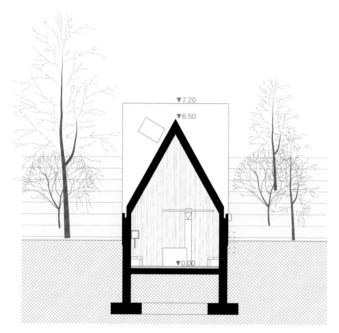

B-B' 剖面图 section B-B'

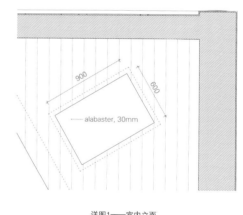

详图1——室内立面
detail 1_interior elevation

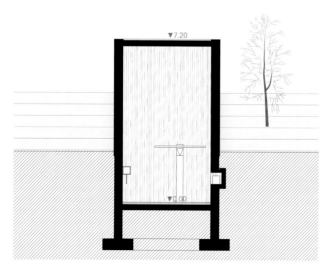

C-C' 剖面图 section C-C'

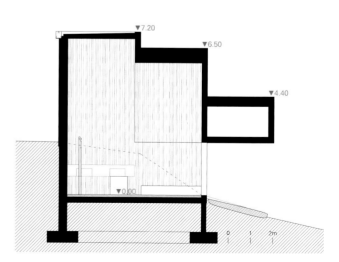

D-D' 剖面图 section D-D'

南美巴哈伊神庙
Bahá'í Temple of South America

Hariri Pontarini Architects

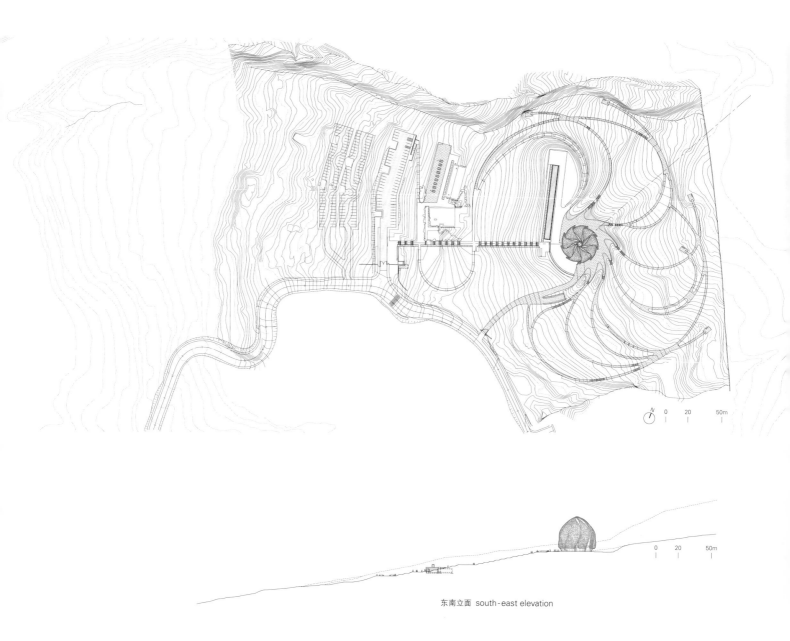

东南立面 south-east elevation

2003年，巴哈伊社区为在南美大陆建造全球第八座也是最后一座神庙举行了一场国际设计竞赛。

神庙坐落在智利圣地亚哥市外的安第斯山脚下，建筑师用光来表达该建筑的精神和设计理念。

在长达十四年的建造过程中，该设计一直致力于表达巴哈伊教的核心，即多样性的统一。该项目的灵感来自巴哈伊教经典："祷告体现在光明中"。设计团队一直专注于创造一种能够捕捉和催化光并能够被光赋予生命的结构的探索。整座建筑只简简单单设计了九个侧面和九个出入口，是一个体现精神层面的结构，是一个开放的场所，欢迎所有怀有不同宗教信仰（或没有信仰）的人们，不分经济阶层，不分文化背景。在巴哈伊教义中，没有神职人员，没有宗教仪式，也没有具有象征意义的图形和符号，这意味着该建筑在许多方面都将是一种全新的表达，无法模仿旧式清真寺、犹太教堂或一般教堂。

在经过对捕捉、表达和体现光线的材质进行深入细致的调查研究后，设计团队开发了两种用于覆盖建筑外立面的材料：一种是来自葡萄牙Estremoz采石场的特殊半透明大理石，用于内层；而外层使用的是专门为本项目研发的压铸玻璃面板。外层浇铸玻璃面板材料由该项目组与多伦多杰夫·古德曼工作室的艺术家们共同合作开发，耗时近四年。1129块独特的压铸玻璃面板，或扁平或弯曲，被精心地组装在一起，成为九个玻璃翼面。

最终，九个完全一样的玻璃翼面优雅地扭转交织在一起，构成一个开放的、易接近的、充满阳光的空间，供人们祈祷和冥想。在黎明和黄昏之间，神庙的玻璃和大理石上便溢满了在圣地亚哥天空中漫舞的季节性色彩。在夜晚，神庙的玻璃和大理石会将庙内的光投射到室外，将柔和的光芒映射到城市边缘的安第斯山脉上。

翼面的超级结构由成百上千个独特的、单独设计的薄型钢构件和节点连接组成。每一个翼面结构都安装在混凝土环和混凝土柱上，而混凝土环和混凝土柱又被建在弹性隔震器上。这样，在发生地震时，混凝土垫就会水平滑动以减缓震动冲击。

该项目的可持续性达到了最高水平，因此对环境的影响很小。圆顶的体量、材料的选择和双壳覆层的设计，都最大限度地发挥了被动供暖和制冷的效果，几乎完全没有供暖和制冷的必要。精挑细选的材

料经久耐用，使神庙能够屹立400年不倒。这座神庙表达了一种包容性的信仰，它不仅仅体现了复杂设计、创新、可持续性发展和建设等方面，也是巴哈伊社区愿望的体现。

In 2003, the Bahá'í community held an international competition for the design of the eighth and final continental Temple for South America.
Set within the foothills of the Andes, just beyond the metropolis of Santiago, Chile, the architects' submission used light for its spiritual and design expression.
Fourteen years in the making, the design sought to express the central tenant of the Bahá'í Faith – unity in diversity. The inspiration for the project was found in the Bahá'í writings:

"a prayer answered is light embodied". The team became immersed in the exploration of creating a structure that would capture, catalyze, and become alive with light.
The design brief simply mandates nine sides and nine entrances, a spiritual structure which is welcoming and open to peoples of all faiths (or no faith at all), all economic strata, and all cultural backgrounds. As there are no clergy, or rituals, or iconography in the Bahá'í Faith, this would, in many respects, be a new expression – one which could not imitate the old forms of mosque, synagogue, or church.
An intensive investigation into the material qualities that capture, express, and embody light resulted in the development of two cladding materials: an interior layer of exceptional translucent marble from the Portuguese

夹层 mezzanine floor

夹层结构 mezzanine structure

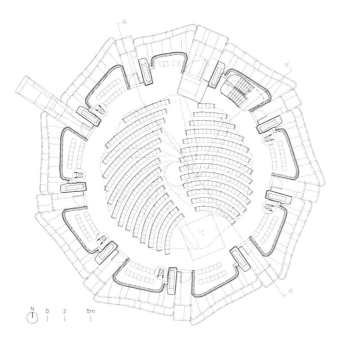
一层 ground floor

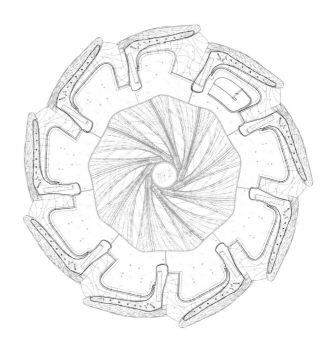
一层结构 ground structure

项目名称：Bahá'í Temple of South America / 地点：Santiago, Chile / 建筑师：Hariri Pontarini Architects
设计团队：Siamak Hariri - Partner in charge / 首席设计建筑师：Justin Huang Ford, Michael Boxer, George Simionopoulos, Tiago Masrour, Tahirih Viveros, Jin-Yi McMillen, Jaegap Chung, Adriana Balen, Mehrdad Tavakkolian, Donald Peters, Jimmy Farrington, John Cook
项目团队：Benkal + Larrain Arquitectos / 当地建筑师：Gartner Steel and Glass GmbH / 超级结构与覆层：Jeff Goodman Studio and CGD Glass / 玻璃覆层：EDM / 石材结构：Juan Grimm / 景观建筑师：Simpson Gumpertz & Heger, Halcrow Yolles, EXP, Patricio Bertholet M. / 结构顾问和机电顾问：MMM Group / 管道顾问：Videla & Asociados / 暖通空调顾问：The OPS Group
照明顾问：Limari Lighting Design Ltda., Isometrix / 声学设计：Verónica Wulf / Way-Finding and Graphics: Entro Communications
用地面积：93,000m² / 建筑面积：1,593m² / 结构：steel / 材料：custom developed cast glass, Portuguese estremoz marble, American walnut, patinated bronze, leather / 造价：30 Million USD / 设计时间：2003—2012 / 施工时间：2012—2016 / 竣工时间：2016.10
摄影师：©Sebastian Wilson Leon (courtesy of the architect) - p.94, p.98~99, p.102, p.103; ©Guy Wenborne (courtesy of the architect) - p.96, p.97[left-top. right-top], p.100[lower]; ©Javier Duhart (courtesy of the architect) - p.97[bottom]; ©Ian David (courtesy of the architect) - p.100[upper]; ©Robert Weinberg (courtesy of the architect) - p.104~105; courtesy of the architect - p.90~91, p.93, p.97[left-middle]

混凝土结构
concrete structure

眼睛结构
oculus

边缘构件
edge members

全部钢材
full steel

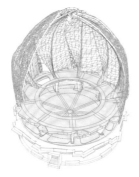
石材
stone

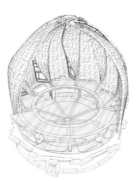
浇铸玻璃
cast glass

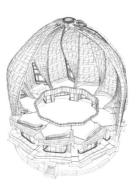
寺庙整体
temple complete

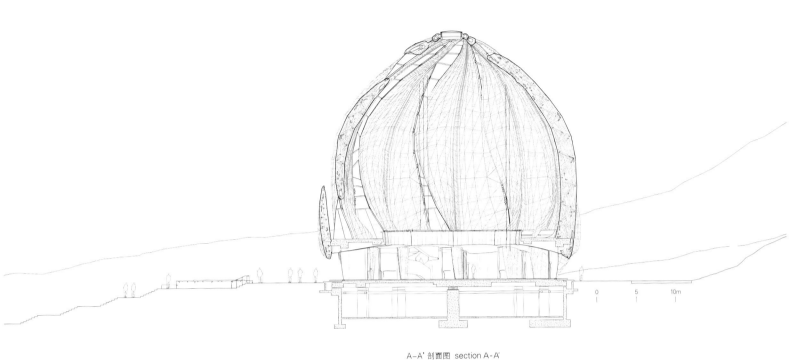

A-A' 剖面图 section A-A'

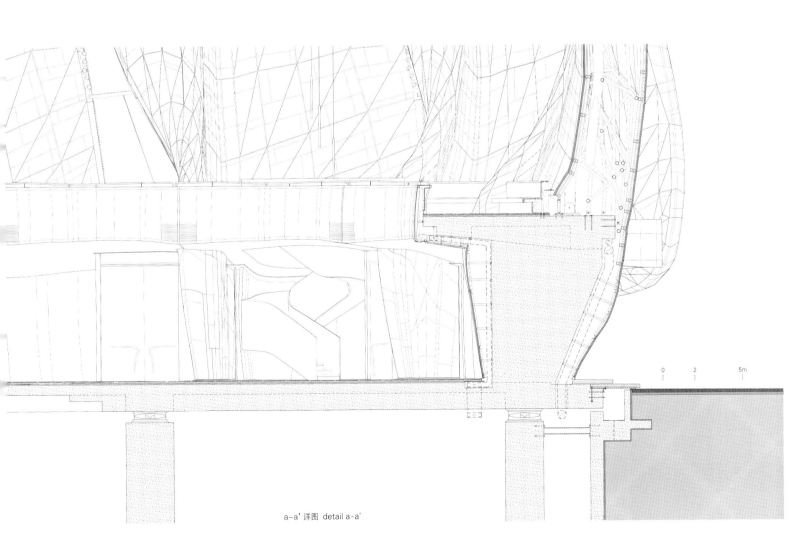

a-a' 详图 detail a-a'

Estremoz quarries, and an exterior layer of cast-glass panels developed exclusively for this project. The research for the cast-glass exterior cladding took nearly four years, working in collaboration with artisans at Jeff Goodman Studio in Toronto. A remarkable 1,129 unique pieces of both flat and curved cast-glass pieces were produced and assembled with meticulous care to create each of the nine wings.

The final design is composed of nine identical, gracefully torqued wings that frame an open, accessible, light-filled space for prayer and meditation. Between dawn and dusk the Temple's glass and marble become infused with seasonal colors that dance across Santiago's sky. At night, the materials allow for an inversion of light, whereby the temple, lit from within, casts a soft glow against the Andean mountain range bordering the city.

The super-structures of the wings are comprised of hundreds of unique, individually engineered slim-profile steel members and nodal connections. Each of the wings rest on concrete rings and columns on elastomeric seismic isolators, so that in the event of an earthquake, the concrete pads slide horizontally to absorb the shock.

The Temple treads lightly on the environment with the project being built to the highest levels of sustainability. The domed massing, materials, and double-shell cladding were designed to maximize the effect of passive heating and cooling almost completely eliminating heating and cooling requirements. Materials were selected for durability and longevity with the Temple being mandated to last 400 years.

Expressing a faith of inclusion, the temple is more than just a story of complex design, innovation, sustainability, and construction; it is the embodiment of a community's aspirations.

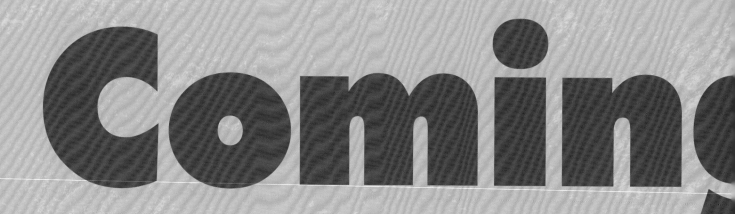

Coming

回家

贾普·道森是研究建筑和身体之间的关系以及建筑对个人和个人宗教信仰的影响方面的专家。作为一名建筑师和教育家,他通过机制来研究宗教建筑空间的重要性。他使用的这些机制是把物理空间和人的行为连接在一起的。

在他的文章中,贾普·道森建议通过内在性来解读宗教空间,去寻找某种只有通过我们自身的内在性才能完全理解的模式(Dawson 2012)。敬思空间需要个人重新与她/他的生理和心理建立联系,而建筑作为一个物理存在,能够维持人的身体和心灵的状态(Dawson 2015)。这是关乎每个人的身体和精神维度的。

在本文中,贾普·道森强调了宗教空间在亲密维度方面的重要性。

Jaap Dawson is an expert of the relationships between architecture and the body, and their impact on the individual and her/his intimate religious beliefs. In his work as an architect and educator, he studied the importance of religious spaces in architecture through the mechanisms that connect the physical spaces and the practice.

With his texts, Jaap Dawson suggests reading religious spaces through the interiority, and to look for patterns that can only entirely be understood through our own interiority (Dawson 2012). Spaces for worship require the individual to reconnect with her/his biological and psychological nature, whereby architecture as a physical presence is able to sustain the body and the soul (Dawson 2015), that are the physical and the spiritual dimensions of every human being.

In his article for this book, Jaap Dawson accentuates the importance of religious spaces as intimate dimensions. Silvio Carta

---

1. Jaap Dawson, *Patterns with a Heart* (Co-editors: Alessia Cerqua, Stefano Serafini, Archana Sharma, Antonio Caperna), 2012: p.97.
2. Jaap Dawson, *Building to Sustain Body and Soul*, BIOURBANISM, 2015: p.49.

Coming Home

Jaap Dawson

狗狗知道哪里最适合它

我八岁的时候,父母终于同意我养一只小狗。在我把它领回家之前,还从附近的废品站找了一个大箱子。我铺了条毛巾在箱子里,把它放在了厨房工作台下面的地板上。对狗狗洛基来说这一定是一个完美的小窝,它会是装在箱子里的小狗狗!

可是实际上,狗狗洛基一点儿也不喜欢这个窝。它在我随便找来的大箱子里待得不舒服,反而钻到了一个刚好能装下它的小箱子里去了。那个小箱子是我以前装生日礼物——玩具卡车的包装盒儿。狗狗洛基是凭着自己的感觉来选择小窝的,它并不会像人类那样思考,但它会用身体去感受待在哪里更舒服。

我八岁时才意识到我得向狗狗洛基学习。它从不胡思乱想,至少不像人类那样瞻前顾后。它对自己的身体十分了解,它知道怎样待着舒服,也知道自己需要什么。

年轻的建筑师不知道如何打造一个适合他的空间

经过多年的建造和绘画学习,经过多年斟酌和尝试各种风格,我终于成为一名建筑师。但我知道什么?我到底知道什么?我知道狗狗洛基仅仅六周大的时候就知道的东西吗?

不,我不知道。我学会复制其他当代建筑师的设计。我被鼓励珍惜当下的美好时代。我将建筑当作技术性的东西,当作功能性的设计布局,认为通过思考,我们可以进行改进和完善这些方面。

The Dog Knows Which Space Fits Him

When I was eight, my parents finally let me have a dog. Before I brought him home, I found a large box that the neighbourhood grocery store had discarded. I placed a towel in it and set it on the kitchen floor under a work table. It would be a perfect nest for Rocky, the Boxer puppy!

But Rocky would have none of it. He didn't feel at all at home in the box I had rigged up for him. Instead, he crept into a box not very much larger than he was, a box that had contained a toy truck sent to me as a birthday gift. Rocky chose with his body. He chose without thinking. He knew where he felt at home.

Even at the age of eight I realized I could learn from Rocky. He didn't think – at least not the way humans think. He knew. He trusted his body and what it needed, what he needed.

The Young Architect Doesn't Know How To Design a Space That Fits Him

After years of building and drawing, of studying and wrestling with styles, I finally became an architect. But what did I know? What did I really know? Did I know what Rocky knew when he was only six weeks old?

难道建筑师不需要了解和知道更多的东西吗？

就这一问题，狗狗洛基做出了回答：它不用思考就知道；它知道哪里有家的感觉；它知道合适的尺寸、恰当的设计，知道什么样的空间才是能为它提供滋养的空间。能为我们提供滋养的空间，这难道不是以基本的方式为我们提供滋养的空间的定义？空间不就应该让我们有家的感觉？空间不就应该让我们表现出真实的自己？

我们究竟是谁呢？

这个问题值得我们思考。我们可以对此抱有疑问，但我们也可以慢慢体会。在我们建造的空间中，我们可以认清自己。在我们所建造的空间里，我们可以体会到初心，体会到本我，体会到这个物质和精神世界里我们生命的存在。哪些空间，哪些建筑，给我们物质和精神世界的体验？又有哪些空间，哪些建筑，让我们有家的感觉？哪些空间，哪些建筑，适合我们？

建筑师发现适合他的空间

我想起了一个我特别喜欢的空间。待在这个空间里，我觉得完全像在家里一样。不只是我的身体感觉像在家里一样，是整个自我。这个空间适合我的身体和灵魂，适合我的身体结构和超越我日常意识的内心世界。在那里，我能感到活着的喜悦；对那里，我的崇拜之情会油然而生。

这就是位于荷兰Schaijk村的一座住宅的主层。它既不是教堂，也不是清真寺或寺庙，也不是为了延续某种传统而建造的。也许正因为如此，它比一座为了让我们更接近某个特定传统而建造的建筑更具有永恒性。

右页图是一张主起居楼层的照片，是一个名叫简·德·琼的建筑师为自己、妻子和五个孩子设计建造的房子。实际上，他建造这座房子是为了检验一个理论，一个由他的导师多姆·汉斯·范·德·拉恩通过多年对现有建筑的实验和分析所得到的比例理论。我现

No. I had been taught to copy the designs of other contemporary architects. I had been encouraged to value the age I lived in as though it were holy. I had learned to regard and conceive buildings as technical things, as functional arrangements, that we could improve and perfect by thinking about them.

Wasn't there anything more that an architect needed to know?

Rocky provided the answer: knowing without thinking; knowing where he felt at home; knowing the right size, the right design, the space that nourished him. The space that nourishes us: Isn't that the definition of a space that feeds us in an essential way? A space that brings us home? A space that fits with who we really are?

Who are we, really?

We can think about it. We can speculate. But we can also experience it. In built spaces we can experience who we really are. In the spaces we build we can experience our built-in nature, our life in both the material and the spiritual world. Which spaces, which buildings, give us the experience of both material and spiritual worlds? Which spaces, which buildings, bring us home? Which spaces, which buildings fit us?

The Architect Discovers a Space that Fits Him

One of my absolute favourite spaces comes to mind. I feel utterly at home in it. It's not only my body that feels at home: it's my whole self. The space fits my body and my soul, my physical frame and the inner world beyond my day-to-day consciousness. The space makes me glad to be alive. The space brings me to worship.

The space is the main level of a house in Schaijk, the Netherlands. It is not a church or a mosque or a temple. It is not a building conceived to fit in an inherited tradition. And that's what makes it ever more timeless than a building built to draw us nearer to a particular tradition.

Here is a picture of the main living level:

The architect, Jan de Jong, built the house for himself, his wife, and his five children. Actually he built it in order to test a theory of proportion which his mentor, Dom Hans van der Laan, had discovered through years of

在说话的口吻像一名思虑过密的建筑师。让我重新回到狗狗洛基的例子上。现在我将谈谈,为什么我感觉这些空间适合我,这些空间给我怎样的感受,以及这些空间让我的意识回归何处。

舒适的实体空间,通往心灵世界的静地

请跟随我的步伐,和我一起漫步在中央空间吧。边走边看,你马上就会明白我们为什么称此处为中央空间了,因为它的两侧各有一排其他空间。这个空间看起来不仅空旷明亮,而且还能给人带来一种安全感。它不像礼堂或机场航站楼那样宽广无限,左边一排的柱子和右边一排的墙体就好像是活的,如舞蹈中两排鲜活的舞者。这支舞蹈的目的是建立一个中心,让这个中心在我们的体验中变得鲜活。

这些"舞者们"是如何将无限的活力与生机带入这个中央空间的呢?就让我们站在这些"舞者"中间来一起揭开谜底吧!我们先从左边的画廊说起,我们可以看到、感觉到并发现,柱与柱之间恰如其分的距离构成了眼前的带有入口的墙体。每两根柱子之间形成一个中心。就这样,左边这排的"舞者们"便创造出了很多较小的中心,与整个活力四射的中央空间相辅相成。

在那排灵动的柱子和后面没有窗户的墙体之间也有一个空间。在这个空间里来回走动是什么感觉呢?先说说我的感觉,再说说我的思考。我的感觉就是,这一空间非常适合我的身体。它的空间尺寸或多或少和柱子(舞者)一样大。当我问自己为什么这一画廊空间感觉如此合适,既不太宽也不太窄时,我注意到空间的宽度就是若干柱子的宽度。大量的元素定义了这一空间,真正地创造了这一空间,也让这一空间充满活力。毕竟,生物是按照自己的尺度来创造空间的。

如果我停下来思考我感觉到的舒适度,我就想到了另外一个中心,那就是起居空间。起居空间一侧是一排鲜活的柱子,另一侧是墙。不用量,凭感觉我就知道墙的厚度

Jan de Jong's house

experiments and analyses of existing buildings. But now I am speaking like an architect who thinks too much. Let me return to the example that Rocky provides. Let me feel how the spaces fit me, what they do with me, where they bring me in terms of my awareness.

The Space that Fits Us Makes Physical Centers that Reflect Our Spiritual Centers

Roam with me through the central space. You know immediately that it's the central space because two rows of spaces on either side of it flank it. The space feels both open and protected. It's not endlessly wide like an auditorium or an airport terminal. It's as though the columns on the left side and the wall segments on the right side were alive, were living beings who arranged themselves in a dance. And the goal of the dance was to establish a center, and to make that center come alive in our experience.

How do they do it, these dancers? Let's answer the question by standing between them. We'll start with the gallery on the left. If we begin with the perforated wall that the columns form by standing at a modest distance from each other, we see and feel and discover a center between every two columns. A row of dancers whose dance makes centers, lines the central space of the living level.

But there's also a space between the row of dancing columns and the windowless wall beyond them. What does it feel like to walk back and forth in this space? First the feeling; then the reflection. My feeling is that the space fits my body, which is more or less the same size as the columns, the built dancers. When I ask myself why the gallery space feels so right – not too wide and not too narrow – I notice that the width of the space is a modest measure of a number of column widths. The massive elements that define the space, that truly give birth to the space, make the space alive too. After all, living beings have made the space according to their own measure.

If I pause to reflect on the fit I feel, then I become aware of another center. The living space is a center between

和柱子的厚度是一样的。墙和柱子共同构建了这样的建筑空间，我的身体在里面感觉非常舒服和自在。在这样的空间里，不用想，我就可以感觉到包容它的更大的空间的尺寸大小。我居住在一个我不仅能感觉到的世界里，也居住在一个我能了解和知道的世界里。

一个我可以了解的世界，一个超出我们传统意识的世界，这难道不是任何建筑或空间要达到的目的吗？了解这样的世界意味着与之建立联系，从中获得精神能量。这座房子还有更多的空间帮助我了解，帮助我建立联系，帮助我让我待在这所房子里感觉舒服自在，有家的感觉。中央空间的右侧是一系列的小房间。我可以独自坐在其中，也可以坐在建筑师妻子或女儿对面与她们聊天。这些房间让我想起了一座恒久不朽的长方形教堂里的小礼拜堂，让我想起了一间简陋的小屋，让我想起了儿时建的林中小屋。它们怎么会让我想到这些的？

墙的厚度使我们可以感受到空间的宽度

我再次从我的感觉、从我的身体感受中回过神来。我看一下这一空间的尺寸，我看一下构成这一简单空间的长宽高，我发现了两个重要元素。这两个重要元素在之前的舞动空间中我就早已发现，就是墙的厚度和一个个小中心。墙的厚度使墙看起来就像一个鲜活的身体，而不是一个抽象的平面。小房间之间的墙体就成为一个个小中心，即房间本身。

如果我只是看构成整个舞动空间的材料表面，那我就看不到这一舞动空间的复杂性。但是，如果结合我对空间的感受和体验，我就能仔细思考这些空间的组成，也就是整个舞蹈的编排。现在，我对这一空间的了解，不只是停留在身体层面，而是更进一步，更深一层。现在，我能完全了解整个空间的设计安排，整个舞蹈的编排。柱子就是一排中心，并共同排列成了一面中间有几个开口的墙。这面墙与正对面没有窗户的墙构成了一个空间中心。而整个单元——有着清晰厚度的墙和墙之间被赋予了生命的空间——构成了更大的中央空间。我们可以感受这个空间，也可以了解这个空间。这一中央空间一侧的一排小房间成为一个个小的中心，与主要空间垂直。

the row of living columns and the wall on its other side. I don't need to see the thickness of that wall in order to assume it's the same as the thickness of the columns. The wall works together with the columns to form a spatial building block. Since my body feels at home in it, I can feel without thinking or reflecting the measure of a larger space as long as the spatial building block is present in it. I'm inhabiting a world I can not only feel: I'm inhabiting a world I can know.

A world I can know: isn't that the goal of any building or space that brings us in contact with a world beyond our conventional awareness? Knowing that world means relating to it, deriving psychic energy from it.

But there are more spaces that help me to know, help me to relate, help me to feel at home in this house. To the right of the central space is a series of tiny rooms. I can sit in them alone or across from the architect's wife or daughter. They remind me of side chapels in a timeless basilica. They remind me of an elementary cabin. They remind me of the houses I built as a boy in the woods. How do they do all that?

The Thickness of the Walls Allows Us to Experience the Width of a Space

Again I pause from what I feel, what my body knows. I take a look at the measures, at the ingredients in this recipe for an elementary space. And I discover two key elements that I've already discovered earlier in this spatial dance: wall thickness and centers. The thickness of the wall segments is like a living body rather than an abstract plane. And the walls between the tiny rooms form centers: the rooms themselves.

If I just look at the surfaces of the materials that form the whole dance, I miss the complexity of the dance. But if I collect my experience of the spaces, I can reflect on the composition, the choreography. Now I know at a deeper level than bodily feeling alone. Now I'm fully aware of the dance of centers. Columns make a row of centers. Together they form a wall with openings. That wall stands opposite the blind wall, forming a spatial center between the two. And the whole unit – the walls with clear thickness and the space they bring to life between them – gives measure to the larger central space. We can know that space as well as feel it. And the row of tiny

经过仔细思考，我再回头看看所有这些空间，我适合这些空间。无论是踱步穿过左侧的画廊，还是在右侧的小型开放空间里独坐或与人促膝而谈，抑或是我自己在主中央空间起舞，我都感觉这些空间是鲜活的，是有生命的。到现在，我已经从左到右，从右到左，体验了这些空间，但我发现，这个中央空间很长，形成了一个中心。一端是桌椅，一端是厨房，中间就是中心，空空荡荡，让人感到很意外（也很庆幸），只有一盏枝形吊灯。吊灯自上而下照亮整个舞动空间。

物理空间可以带领我们走向精神空间：我们的家

现在是时候反思我的经历了，是时候挖掘我为什么选择这所房子作为我与自然接触的建筑的最好例子了。理由很简单。不管我们继承了什么故事和仪式，不管我们怀揣怎样的梦想，精神上我们都以一个中心为中心。这个中心可以是一幅沙画的中心，可以是一颗恒星的中心，可以是喷泉，可以是复杂的瓷砖设计。最重要的一点，也是显而易见的事实，那就是这个中心就是我们自己的心。从心理学上来说，是无意识的心产生了我们的自我。但是，如果我们的自我忘记了生它的心，那我们的自我就不会真正活过，就不会自由自在。

我差点忘了，我不是放在箱子里的狗，我也不只是一个在一座特别的房子里会感到无拘无束的人，我也是一名建筑师，也就是说我已经吸取了教训。在我的设计里，我不想只看形式或风格。实际上，我不想看，我想去感受。我想去感受大量的材料可以为我和他人创造出怎样充满活力的空间。我想——不，我感受到自己需要，去设计一个可以让我接触到我自己的中心，接触到生活本身的中心。我想设计出能吸引我回家的建筑。

rooms on the other side of the central space gives us centers perpendicular to the main space.
After reflecting, I return to all the spaces. I fit in them. I experience them as living, as alive, whether I walk through the gallery on the left, whether I sit and talk in one of the tiny open rooms on the right, whether I myself dance through the main central space. Till now I've experienced the spaces from left to right and then back again. But now I discover that the central space gives me a center in its long length. At one end stands a table with chairs. At the other end the kitchen. In the middle is a center, unexpectedly (and thankfully) empty, save for a chandelier. The chandelier sheds light on the vertical dimension of the dance.

A Physical Space Can Bring Us in Contact With A Spiritual Space: Our Home
Now it's time to reflect on my experience: now it's time to discover why I choose this house as a prime example of a building that puts me in touch with nature. And the reason is simple. Regardless of the stories and rituals we inherit, and in addition to the dreams we receive, we all revolve psychically around a center. That center may be the heart of a sand painting. It may be the middle of a star. It may be a fountain. It may be an intricate tile design. The essential point, the clear truth, is that the center is our own heart. Psychologically it's the unconscious heart that gives rise to our ego. But our ego doesn't really live, doesn't really feel at home, if it forgets the heart that gave birth to it.
I almost forgot: I'm not a Boxer dog. Neither am I just a man who feels at home in a particular house. I'm also an architect. And that means I've learned a lesson. In what I design, I don't want to look at forms or styles alone. In fact, I don't want to look. I want to feel. I want to feel what massive materials can do to create spaces that come alive for me and for other people. I want – no, I feel the need in myself – to design centers that put me in touch with my own center, with the center of life itself. I want to design buildings that invite me to come home.

白俄罗斯纪念礼拜堂
Belarusian Memorial Chapel
Spheron Architects

白俄罗斯纪念礼拜堂是自1666年大火以来在伦敦建造的第一座木质教堂。该建筑由Spheron建筑师事务所设计,选址于Woodside公园。它是为英国的白俄罗斯侨民社区建造的,旨在纪念1986年切尔诺贝利核灾难的受难者。核反应堆爆炸对白俄罗斯造成的影响尤其严重,其中70%的放射性尘埃降落在白俄罗斯,迫使成千上万的白俄罗斯人离开家园并重新定居在包括英国在内的世界各地。

白俄罗斯纪念礼拜堂坐落在伦敦北部的一个白俄罗斯人社区和文化中心,这里被称为Marian House,周围环绕着17棵受保护的树木。该礼拜堂像白俄罗斯的许多乡村教堂一样,毫不突兀地与周围的花园树木融合在一起。

该礼拜堂的设计经过了对白俄罗斯木质教堂传统的精心研究。Spheron建筑师事务所创始人、主管Tszwai So曾花费大量时间在白俄罗斯记录和绘制传统教堂,其中包括自三十年前切尔诺贝利灾难以来所废弃的村庄上的一些建筑。圆顶尖塔和木瓦屋顶是白俄罗斯数百座传统教堂的共同特征。这些特征可以让身处伦敦的白俄罗斯人感到亲切和慰籍,带给他们回忆。他们其中很多人在切尔诺贝利灾难后迁居英国,而其他人在随后的国内政治经济动荡中被迫搬离故乡。

不过,Spheron建筑师事务所还是在基本的传统形式中引入了一系列当代设计,例如,侧壁波浪般起伏的木质褶边,它让建筑外观更加生动活泼。自然光线通过礼拜堂长轴方向靠近地面部分隐蔽的窗户以及前立面装有磨砂玻璃的高大窗户进入室内。晚上,礼拜堂内部柔和的光线让礼拜堂微微发光。礼拜堂内部有一个木质屏风,上面装饰有一系列历史图标。这道屏风将礼拜堂中殿与后殿中的圣坛区隔离开来。

这个69m²的小礼拜堂由罗马教廷资助,将取代白俄罗斯天主教会在现有社区中心内的临时礼拜场所。这座新礼拜堂可容纳多达40人,它不仅成为白俄罗斯社区重要的精神聚集地,也是1986年切尔诺贝利灾难受难者的永久纪念碑。

该建筑饱含着白俄罗斯历史和文化的象征意义,同时也深深融入了英国的建筑创新。这座小礼拜堂是白俄罗斯社区面对灾难时有形的精神象征,一线希望和信仰表达。尽管该礼拜堂是当地白俄罗斯人欣赏和利用的主要场所,但其目标是让所有拥有好奇心的人来参观,并激发人们对从中汲取艺术灵感的那个国家(白俄罗斯)的兴趣。

The Belarusian Memorial Chapel is the first wooden church built in London since the Great Fire of 1666. Designed by Spheron Architects, the chapel in Woodside Park has been built for the Belarusian diaspora community in the UK, and is dedicated to the memory of victims of the 1986 Chernobyl nuclear disaster. The after-effects of the nuclear reactor explosion were felt particularly severely in Belarus, where 70% of the fallout fell, forcing thousands of people to leave their homes and resettle around the world, including in the UK.

The chapel sits surrounded by 17 protected trees in the grounds of Marian House, a community and cultural center for the UK Belarusian community in north London. Like many rural churches in Belarus, the chapel will offer a gentle presence among the trees of its garden setting.

The chapel was designed by following painstaking research into Belarus's wooden church tradition. Spheron Architects' Tszwai So, a founding director, spent time in Belarus record-

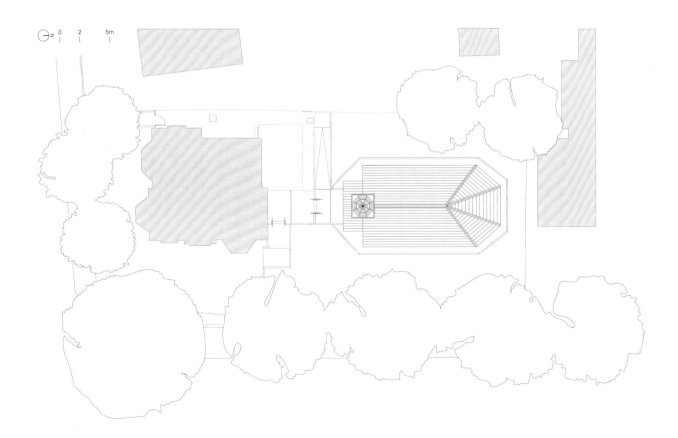

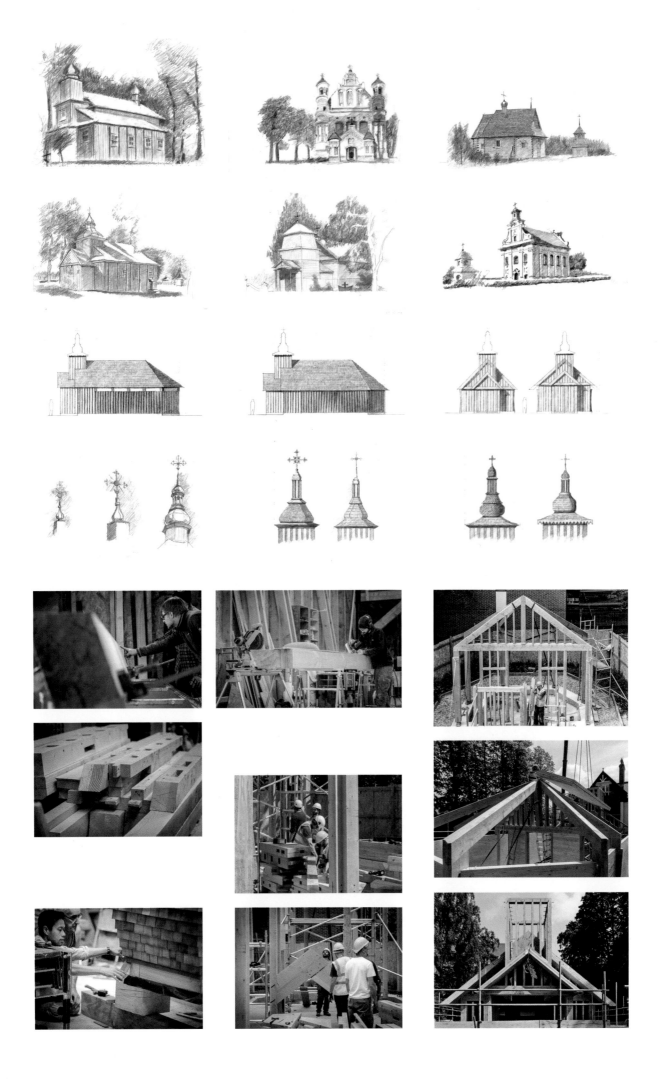

117

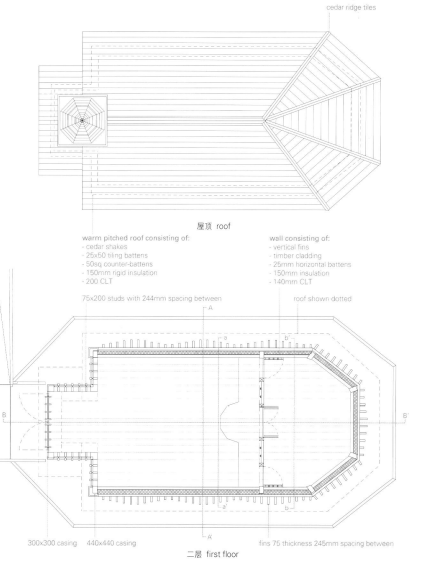

屋顶 roof

warm pitched roof consisting of:
- cedar shakes
- 25x50 tiling battens
- 50sq counter-battens
- 150mm rigid insulation
- 200 CLT

wall consisting of:
- vertical fins
- timber cladding
- 25mm horizontal battens
- 150mm insulation
- 140mm CLT

cedar ridge tiles

75x200 studs with 244mm spacing between

roof shown dotted

300x300 casing   440x440 casing   fins 75 thickness 245mm spacing between

二层 first floor

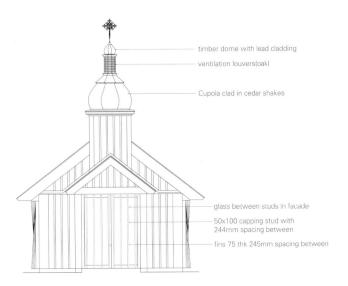

项目名称：Belarusian Memorial Chapel
地点：Woodside Park, London
项目建筑师：Tszwai So, Samuel Bentil-Mensah - Spheron Architects
项目经理：Diocese of Westminster
CDM协调人：BBS SITE Service LLP
机械设备工程师：Arup
结构工程师/总承包商：Timberwright Ltd
规划顾问：Alpha Planning
工料测量师：Change Project Consulting
总楼面面积：75m²
设计时间：2012—2015 / 施工时间：2015—2016 / 竣工时间：2017
摄影师：
©Helene Binet (courtesy of the architect) - p.112~113, p.115, p.122, p.123
©Joakim Boren (courtesy of the architect) - p.117, p.119, p.120

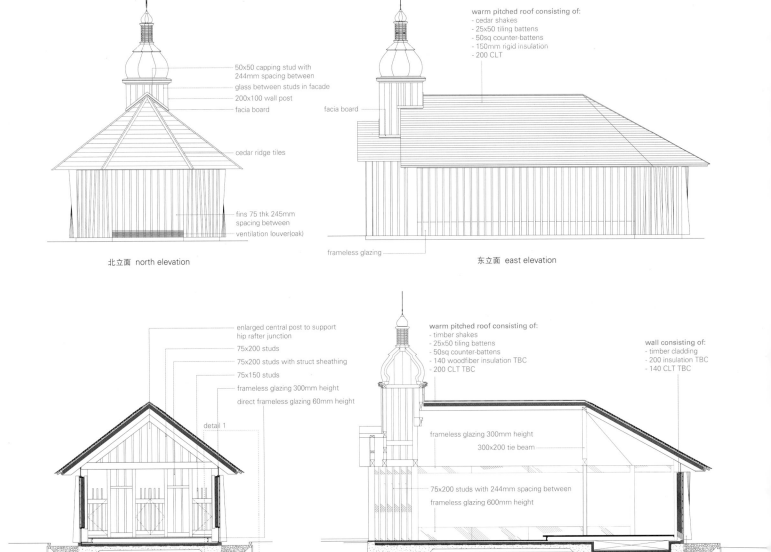

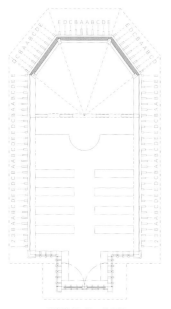

翼板型材与配置
fins profiles and configuration

翼板位置,供90块翼板
fins position
in total 90 fins

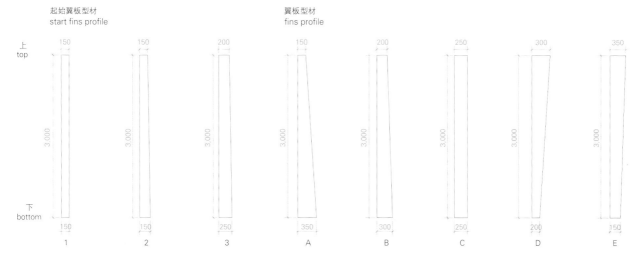

起始翼板型材
start fins profile

翼板型材
fins profile

上 top

下 bottom

| 1 | 2 | 3 | A | B | C | D | E |

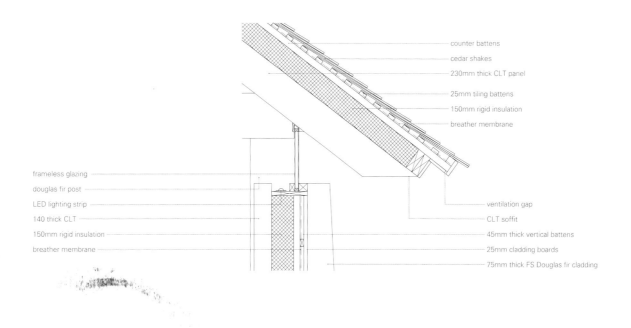

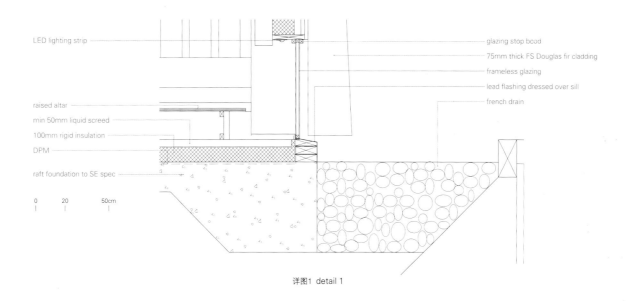

详图1 detail 1

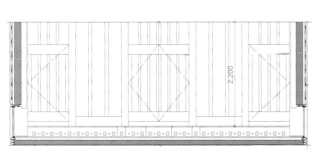

a-a' 详图 detail a-a'

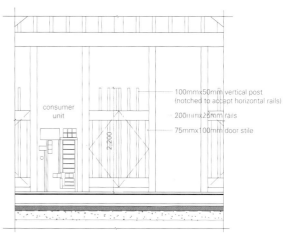

b-b' 详图 detail b-b'

ing and sketching traditional churches, including buildings in villages that had lain abandoned since the Chernobyl disaster thirty years ago. The domed spire and timber shingle roof are common features of hundreds of traditional churches in Belarus and will offer familiarity, comfort and memories to London's Belarusian community, many of whom moved to the UK following the Chernobyl disaster, while others have displaced by subsequent political and economic upheaval in their homeland.

Spheron Architects have introduced a series of contemporary twists to the basic traditional form, however, such as the undulating timber frill of the flank walls which enlivens the exterior. Natural light enters through low-level and concealed clerestory windows running the length of the chapel, and through tall frosted windows on the front elevation. At night, soft light from within allows the chapel to gently glow. Inside the chapel will be decorated with a series of historic icons set into a timber screen separating the nave from the altar area in the apse.

The 69 sq.m. chapel has been funded by the Holy See, and replaces the Belarusian Catholic Mission's makeshift place of worship inside the existing community center. Accommodating up to 40 people, the new chapel serves not only as an important spiritual focus for the Belarusian community, but also as a lasting memorial to the victims of the 1986 Chernobyl disaster.

This project is saturated in Belarusian historical and cultural symbolism, while also firmly embedded in UK architectural innovation. This chapel stands as a visible spiritual token, a ray of hope and an expression of faith by the Belarusian community in the face of disaster. Whilst being a focal point for Belarusian locals to enjoy and utilise, the chapel aims to be viewed with welcoming curiosity by everyone, and encourage interest in that country from which its artistic inspiration was drawn.

圣温塞斯拉斯教堂
St. Wenceslas Church

Atelier Štěpán

位于萨佐维采这个小村庄的圣温塞斯拉斯教堂是由摩拉维亚地区的Atelier Štěpán建筑事务所设计的一座现代圆形建筑。当地居民关于在萨佐维采修建教堂的想法由来已久,从两次世界大战之间时期就开始了,到了2011年才又一次被提上议事日程,并成立了教堂建设委员会。

首要任务是选址,寻找一个可以凸显教堂神圣地位的合适位置。Atelier Štěpán找到了四处基本适合建教堂的位置,其中一处位于萨佐维采村的中心位置,与周围建筑完美融合,同时可以与当地社区和谐相融。

周围的建筑围绕着教堂,如同一个坚实的港湾,凸显了教堂的重要。教堂的形式始于一个简单的圆柱体,不仅与港湾的形状完美契合,同时圆也是一个神圣的符号,通常用来象征神性(与矩形相反,通常象征着世俗)。教堂位于村庄的中心,位于主干道的交会处,彰显教堂所在位置的神圣。

Atelier Štěpán的目标是赋予这座建筑一种幻灭感。墙体两端逐渐变薄变细,营造出一种失重感。视觉上,教堂外观如同一个被切开的纸筒,准备去开始探索空间和体量的新可能性。窗户的设计看起来就像被轻轻推入和拉出纸筒的纸片,让柔和的光线滑入教堂。

驱动该设计的理念是对人产生影响:该教堂的设计旨在有意或无意地影响人的内心。来到教堂里的人将会在这个特定的空间里感受到神性的存在,感受到教堂墙外超然力量的存在。教堂意在引导人们进入,给踏入者带来宁静平和之感,为个体提供与上帝单独对话的私密空间,使每个人都可以亲身体验。

教堂内部装饰简约,充满诗情画意。传统教堂里总是充满了通过绘画、雕塑和装饰品传达的视觉信息。例如,巴洛克时期教堂的室内空间布满了叙事性的绘画,目的是让那些目不识丁的礼拜者得以了解耶稣的生平信息。但在如今信息爆炸的时代,人们更倾向于寻找一个安静的冥想之地以审视内心。这就要求教堂在一定程度上与外界隔开。圣温塞斯拉斯教堂就遵循着这样的设计原则,从窗户进入的光只直接照射到司祭席(译者注:教堂内位于教堂东侧唱诗班席之后的神职人员座席)和圣坛上。圆柱形结构让人联想到一个安全的避难所,待在里面的人可以仰视并找到上方的天窗。天窗从底部开始呈三角形,随着高度的升高而变化,最终变成一个圆。

圣坛以青铜饰面,呈简单的有机形状。其光洁的表面象征着上帝的触摸(因米开朗基罗作为西斯廷教堂天顶的壁画而闻名)。在设计过程中,建筑师探索了弥撒仪式之间的联系,这和人与人之间的联系非常相似。触摸创造运动,引发新的运动。在切线和圆接触的那个奇异点,能量从一次弥撒传递到另一次弥撒,连续不断地传递着一个信息。这让人想起坚信礼时举行的仪式,主教对信徒成圣时抚头顶祝福,也让人联想到自耶稣降临以来延续了两千年的信息传递链条。

The Church of St. Wenceslas in Sazovice is a modern rotunda designed by Moravian architectural office Atelier Štěpán. There had been much discussion of building a church in Sazovice since the interwar period, with locals proposing its construction again in 2011 and founding a church construction association.

The first and most important task the architects faced was to find a location that would lend the church a spiritual atmosphere. Atelier Štěpán identified four possible sites that fit the outlined criteria; one in particular was located at the heart of Sazovice, perfectly aligned with the surrounding structures and lending itself well to harmony with the local community.

An array of structures forms a permanent bay for the church to fit into, emphasizing the church's importance. The church building was originally envisioned as a simple cylinder, not only to fit the shape of the bay, but also to utilize the circle motif, which often symbolizes divinity (in contrast to the rectangle, a symbol of worldliness). The church stands at the heart of the village, directly at the main intersection, asserting its presence as a sacred site.

Atelier Štěpán's aim was to imbue the building with a sense of evanescence. The walls taper off at the ends, giving it a sense of weightlessness; visually, it resembles a paper cylinder being cut open, poised to start exploring new possibilities of space and volume. The windows are designed to look like paper flaps being gently pushed into and pulled out of the cylinder while letting in soft rays of light that glide into the church.

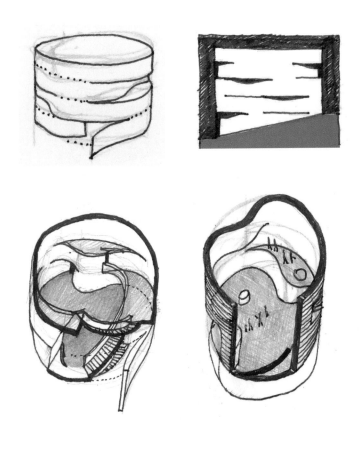
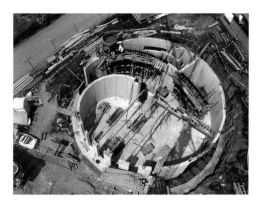

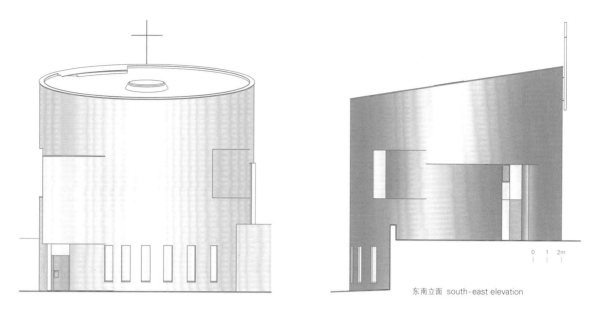

西南立面 south-west elevation

东南立面 south-east elevation

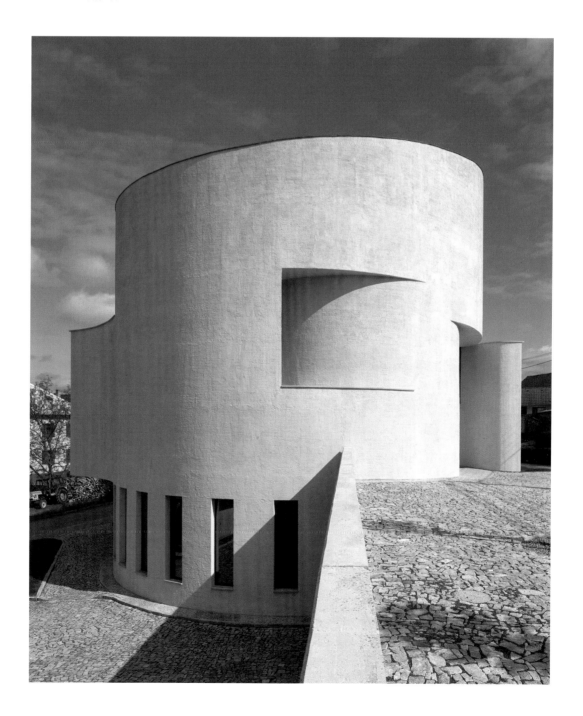

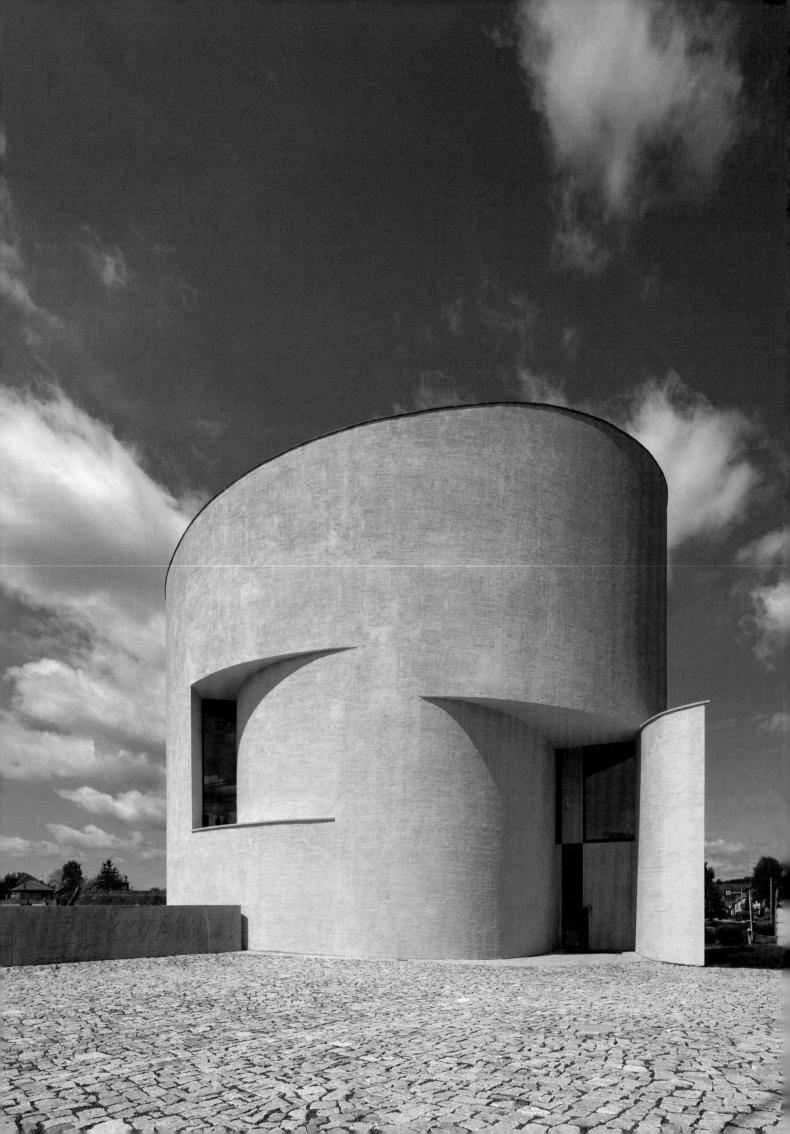

项目名称：St. Wenceslas Church / 地点：Sazovice, Czech Republic / 建筑师：Marek Štěpán - Atelier Štěpán
合作者：Jan Vodička, František Brychta, Jan Martínek, Tomáš Jurák, Hana Kristková, Vladimír Kokolia
承包商：Stavad s.r.o. / 客户：St. Wenceslas Church Building Association / 建筑面积：total floor area 450m²; usable floor area 371m²; building footstep 186m² / 结构：reinforced in situ cast concrete with masonry - walls; in situ cast reinforced concrete - floor slabs; wooden trusses - roof / 材料：concrete, masonry, wood, glass, steel, brass (altar, ambon, sanctuarium)
造价：cca 2 mil. USD / 设计时间：2012—2015 / 施工时间：2015—2017
摄影师：©Jakub Skokan, Martin Tůma - BoysPlayNice (courtesy of the architect)

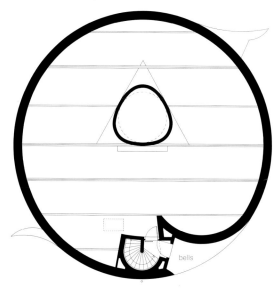

四层 third floor

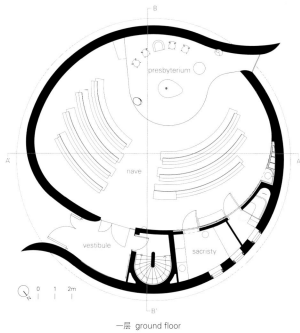

一层 ground floor

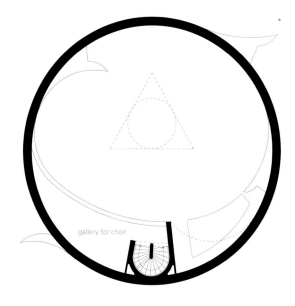

三层 second floor

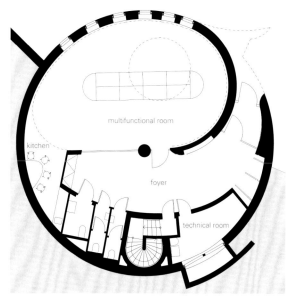

地下一层 first floor below ground

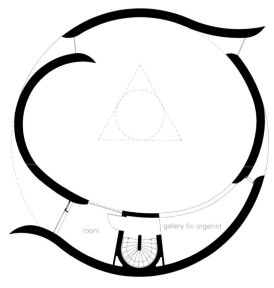

二层 first floor

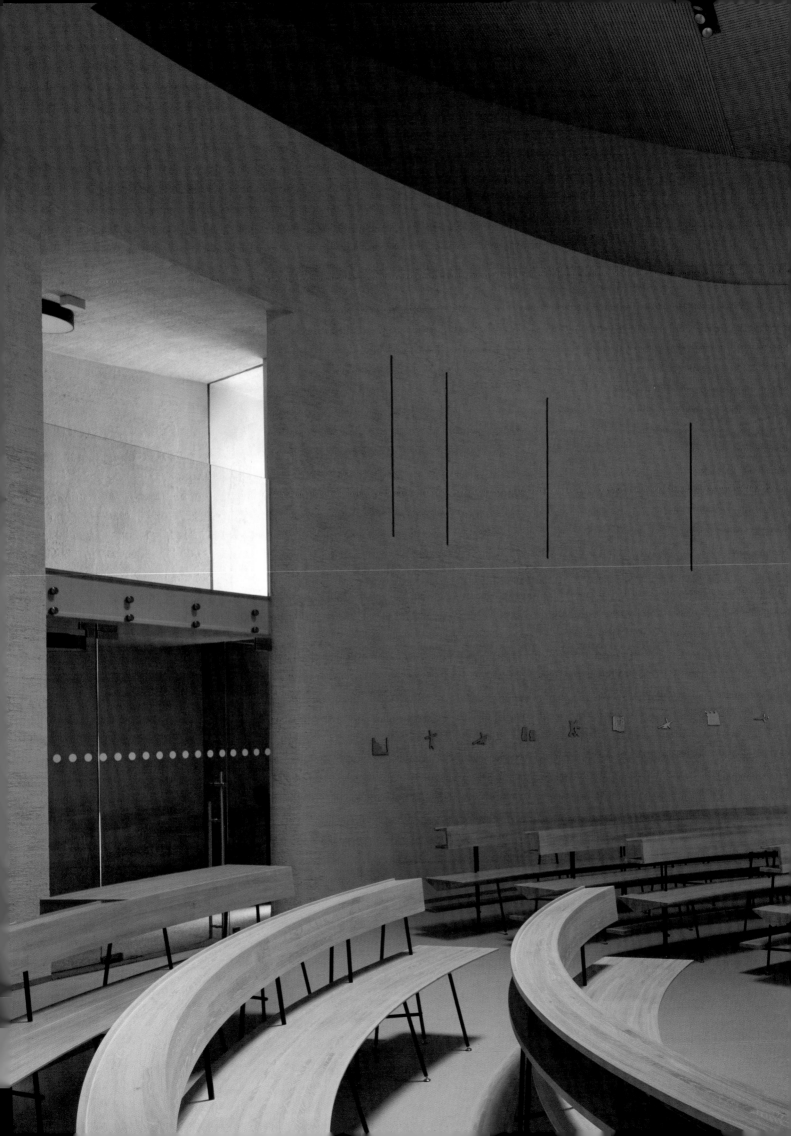

The driving principle behind this work is the idea of having an impact on people; the church is designed to influence people's hearts, whether implicitly or explicitly. Visitors in the structure will feel a hint of the divine in the defined space and the presence beyond its walls. The church is meant to be inviting, promising peace and quiet to those who step inside. It provides a personal environment for the individual to commune alone with God, and is meant to be experienced in person.

The interior is poetically modest with minimalistic decorations. Churches have traditionally been full of visual information conveyed through paintings, sculptures, and ornaments – Baroque-era churches, for instance, were specifically designed with illiterate worshipers in mind, providing information about the life of Jesus in a way anyone could understand. Today, however, people are bombarded with information and seek places of stillness and quiet, where they may meditate and look inwards. This requires a degree of isolation from the outside, which the design of this church provides by allowing light from the windows to directly reach only the presbytery and the altar. The cylindrical structure is evocative of a safe shelter, where the occupant may look upwards and find the skylight above. The skylight starts as a triangle at the bottom and transforms as it rises higher, ultimately becoming a circle.

The altar is a large bronze shell wrought in a simple, organic shape. The pristine surface symbolizes God's touch (as made famous by the Michelangelo fresco at the Sistine Chapel). During the design process, Atelier Štěpán explored the idea of contact between masses, which is strikingly similar to contact between people. Touch creates movement, kicking off new motion. At that singular point of contact between tangent and circle, energy is transferred from one mass to another in a connection that conveys a message. It is reminiscent of the rituals performed during confirmation, the laying of hands on believers during sanctification, and the passing down of messages in a chain that has continued for the past two millennia since the coming of Jesus.

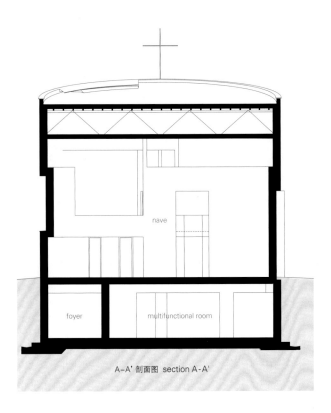

A-A' 剖面图 section A-A'

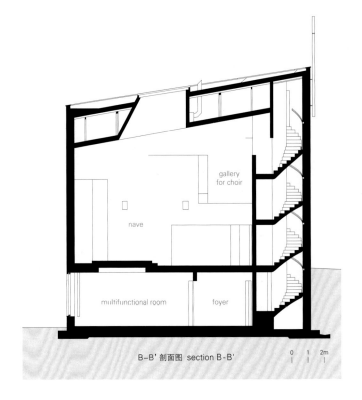

B-B' 剖面图 section B-B'

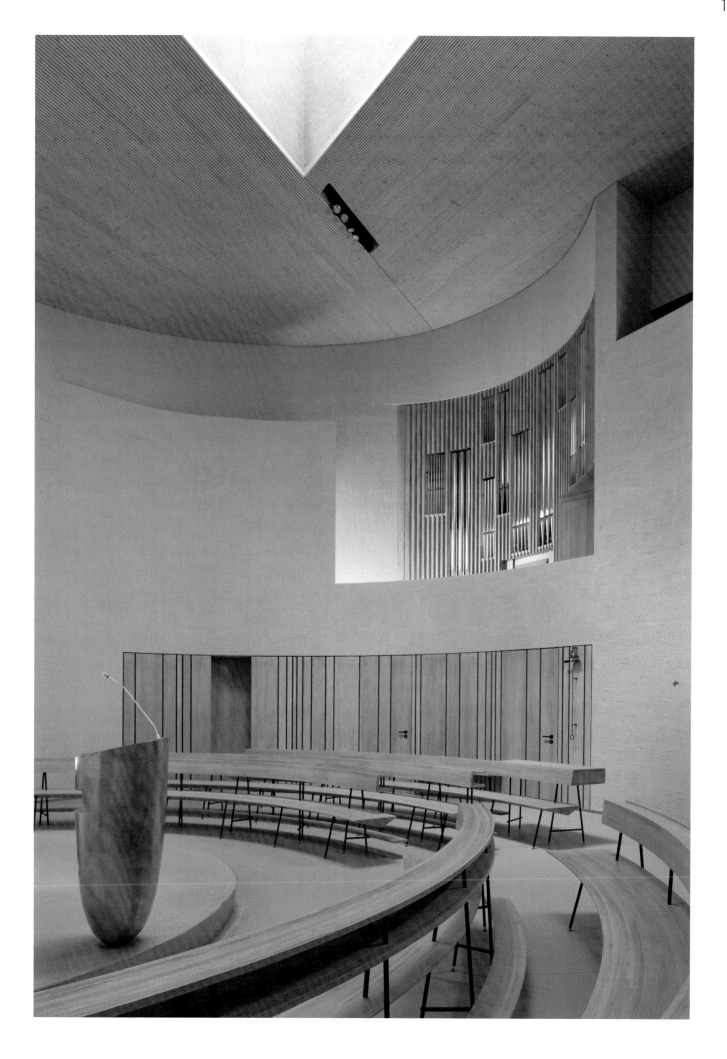

阿格里礼拜堂
Agri Chapel

Yu Momoeda Architecture Office

这是一座带有分形结构系统的日式木质小礼拜堂。

这座小礼拜堂矗立在长崎市一个住宿和培训设施的前面。2016年4月，其管理权转给了一家私人公司，被改造为一家带有婚礼大厅的酒店。酒店需要一个多功能厅，用作新人结婚或者其他庆典活动的场地。场地周围是一座俯瞰大海的大型国家公园。百枝建筑事务所试图将礼拜堂活动与自然环境无缝连接起来。

在长崎，有一座日本最古老的木质哥特式礼拜堂，叫作"大浦天主礼拜堂"。这座小礼拜堂不仅是一个著名的旅游景点，也是一个广受市民喜爱和关注的地方。建筑师通过日式木结构系统把本案这座建筑设计为一座新哥特式礼拜堂。典型的哥特式礼拜堂通常分为三层，分别为"大拱廊""三叶草"和"高侧灯"。该礼拜堂设计采用类似的结构单元，按三种不同尺度分层：基础设施、建筑和产品。多个层次构成了建筑现在的高度，顶部和底部有不同的比例。

网格状平面在每层旋转45°，单元构件或缩小变为原来的1/2，或增加一倍。因此，地面一层有四个大支柱，天花板由16个小支柱支撑，这是一个合理的空间和结构解决方案。

木质结构的空隙形成一个穹顶一样的坚固几何形状，跟圣索菲亚大教堂的穹顶一样。该结构还形成一个中殿和侧面走廊。

一束束向上延伸的树状单元通过缩小*和增大，营造了一个穹顶。第一层柱子单元由4根120mm的方柱组成，第二层由8根90mm的方柱组成，最高一层由16根60mm的方柱组成。建筑师通过减少最底层树状单元的数量来增大可使用面积。这些树状单元利用传统日式木结构系统建造。建筑的四角承重墙承担了水平受力，而内部的木结构则承受约25t的屋面荷载。

*缩小1/2，旋转45°

This is a Japanese-wooden chapel with a fractal structure system.

This chapel stands in front of an accommodation and training facility owned by Nagasaki city. On April 2016, the managing right shifted to a private company, and changed it to a hotel with a wedding hall. They needed a multipurpose hall for the hotel and wedding ceremonies. The site is surrounded by a large national park overlooking the sea. Yu Momoeda tried to connect the activities of the chapel to the natural surroundings seamlessly.

In Nagasaki, there is an oldest wooden gothic chapel in Japan known as "Ohura-Tenshudou". This chapel is not only a famous tourist point, but a place loved and cared by the

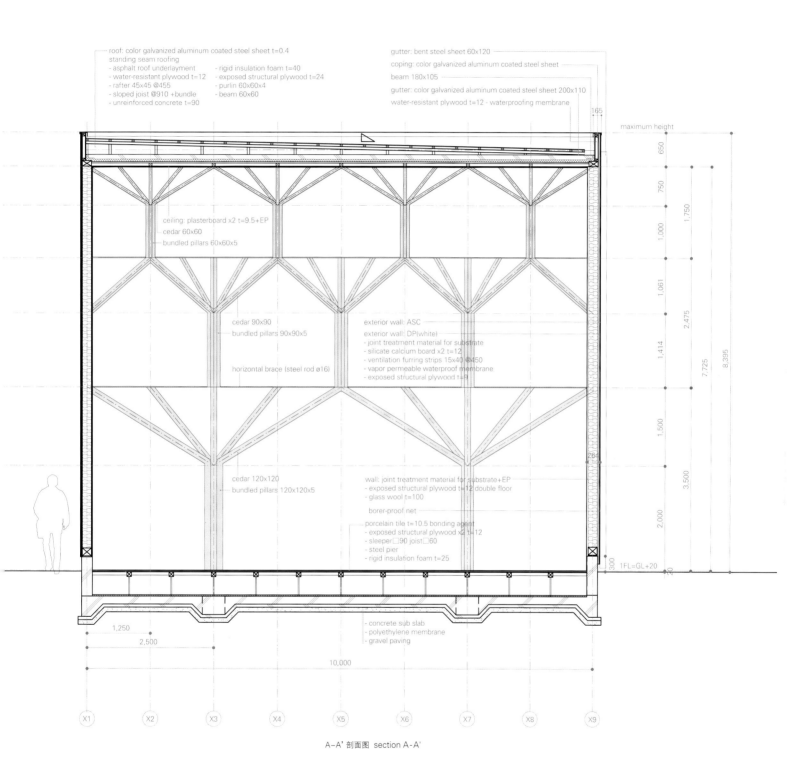

A-A' 剖面图 section A-A'

项目名称：Agri Chapel / 地点：Nagasaki, Japan / 建筑师：Yu Momoeda, Yuko Abe, Takayo Fuchigami-Yu Momoeda Architecture Office / 结构工程师：Mika Araki-Jun Sato Structural Engineers Co., Ltd./ 暖通空调工程师：Ittetsu Koga, Masaru Murayama-Koga Sekkeishitsu / 照明设计：Masaaki Sato, Ryohei Koyama, Tatsuya Fujii-ModuleX / 施工：Yuji Ide, Masanobu Ide, Yoshihiro Iwanaga-Yushin Construction / 家具：Takaya Ishikawa-AURA CREATE / 客户：Memolead / Use: chapel
用地面积：7,927.55m² / 建筑面积：125.27m² / 总楼面面积：125.27m² / 屋檐高度：7,745mm / 屋顶高度：8,395mm / 结构：wood / 室外饰面：long durable for steel structures paint finish(white) over joint treatment material for substrate (wall), color galvanized aluminum coated steel sheet t=0.4 standing seam roofing (roof)-Chapel; porcelain tile □600 t=10.5 (floor)-Corridor; troweled concrete (floor)-Outdoor unit depot; Japanese lawn grass (floor)-Garden / 室内饰面：porcelain tile □600 t=10.5 bonding agent (floor), emulsion paint finish(white) over joint treatment material for substrate (wall), double plasterboard t=9.5 emulsion paint finish (ceiling), synthetic oil paint finish, wood protection finish-Chapel / 设计时间：2015.11—2016.5 / 施工时间：2016.6—2016.10 / 摄影师：©Yousuke Harigane (courtesy of the architect)

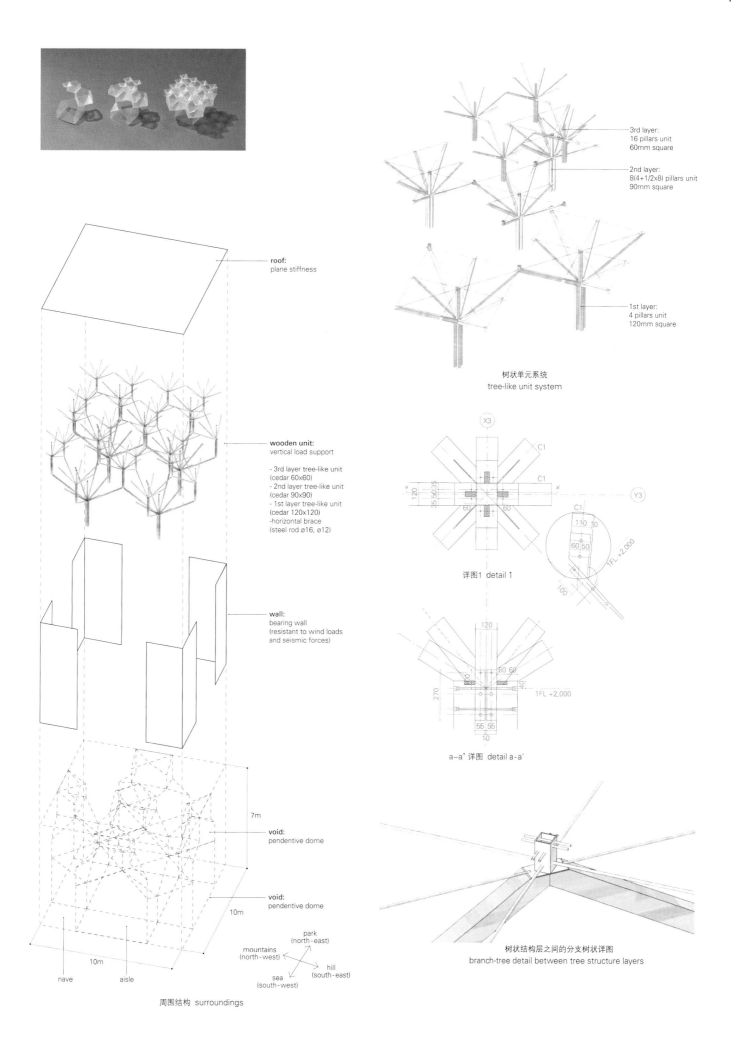

townsfolk. The architect designed the building as a new gothic style chapel, by using Japanese wooden system. A typical gothic chapel is divided to three layers, known as "big arcade", "trifolium" and "high side light". Similar structure unit was layered in three different scale; infrastructure, architecture, and product. The height stands out by the layers, and the top and bottom has different scale experience. The grid plan rotates 45° in every layer. The unit becomes 1/2 smaller, and doubles. As a result, the ground level has 4 big pillars, and the ceiling is supported by 16 small pillars, which is a reasonable solution both to space and to structure.

The void of the wooden structure has a solid geometry like a pendentive dome, just like the Hagia Sophia Cathedral. The structure also makes a nave and side corridor.

A pendentive dome was created by piling up a tree-like unit that extends upward by shrinking* and increasing. Starting by four 120mm square pillars units, the second layer is composed by eight 90mm square pillars units, and the last layer by sixteen 60mm square pillars units. The architect could provide usable open space by reducing the pillars near floor level. These tree-like units are constructed by Japanese wooden system. The four corner bearing walls undertake the horizontal force, and the inner wooden unit supports roof load which count up to 25 tons.

* Shrinking by 1/2, rotating 45°

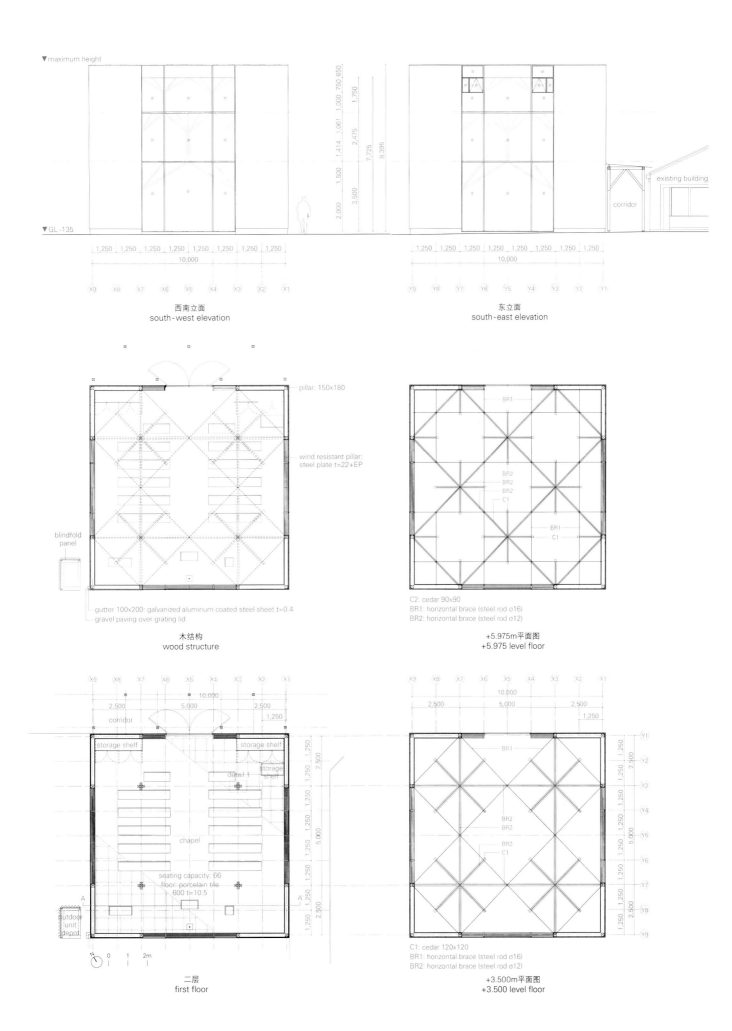

苏州礼拜堂
Suzhou Chapel
Neri&Hu Design and Research Office

这座小礼拜堂是一个较大"村落空间"里的特色建筑。正因如此，它的地理位置得天独厚，不论是从主干道路上还是从滨水区域都可以看到它。它的建筑语言和项目中其他地方所采用的建筑元素大同小异，例如，起伏堆砌的砖墙和灵动的白色体量。但是，这些元素在这里被组合到一起又是那么浑然天成，达到了另一个层次。砖墙被精妙地分解成一个个小墙体，呈现出不同的高度和层次，看起来错落有致，创造出精心设计的景观之旅，引导人们进入该建筑。

白色体量也是经过特殊处理的，分为内外两层。内层就像一个简单的盒子，四周墙体不规则地点缀着窗户。而外层则是一排排金属栅栏，如同一层面纱遮挡着里面的实体体量。白天，白色的盒子在阳光的照耀下闪闪发光，里面的构造隐约可见。夜晚，白色的盒子又像宝石般的灯塔，灯光透过窗户向四面八方散发出柔和的光芒。

在建筑内部，访客首先会被引入前功能区，然后进入主要的礼拜空间。这是一个充满阳光的12m高的空间，与周围的自然景色天衣无缝地融为一体，人们可以透过窗户看到外面的风景，就好像一幅画挂在建筑内。夹层位于礼拜堂大厅上方，设有额外的座位，能容纳更多的人。夹层还有环绕整个空间的狭窄人行通道，可以360°环视礼拜大厅内部。另外，夹层被结合进木质百叶笼子元素中，这个笼子环绕包裹着整个上部空间。网状分布的悬垂灯泡和精致的青铜元素给这个原本宁静的修道院空间增添了一种富丽堂皇的感觉。定制的木质家具，精雕细琢的木工细节与灰砖，水磨石和混凝土等简单材料相得益彰。

该礼拜堂建筑的另一个特点是主空间旁边有一部单独的楼梯，访客可以直接上到屋顶，俯瞰湖面上无与伦比的风景。楼梯的两侧设计了许多开窗，为人们提供了意想不到的室内外景观。

The chapel is a feature building within the larger Village zone. As such, it occupies a prime location visible from the main road and along the waterfront. Its architectural language is derived from similar elements found elsewhere in the project, such as the undulating brick walls and floating white volume – but they are here, taken to another level of articulation. The brick walls begin to break down to an even more refined scale, where different heights of walls interweave with each other to create a choreographed landscape journey leading into the building itself.

The white volume also receives special treatment, here, it is composed of two layers. The inner layer is a simple box punctuated on all sides with scattered windows, while the outer layer is a folded and perforated metal skin, a "veil"

项目名称：Suzhou Chapel / 地点：199 Yangchenghuan Road, Yangcheng Lake, Suzhou, China
建筑师与室内设计师：Neri & Hu Design and Research Office
设计团队：Lyndon Neri, Rossana Hu – Founding partners, Principal in charge; Nellie Yang – Senior associate; Jerry Guo – Senior architectural designer & Project manager; Begona Sebastian – Senior architectural designer; Shirley Hsu; Dana Wu; Maia Peck; Brian Lo – Senior associate, product design; Simin Qiu
用途：chapel and event space / 总楼面面积：700m² / 材料：recycled gray brick, concrete, white perforated metal, white plaster – Architectural materials; recycled gray brick, concrete, terrazzo, white oak, bronze – Interiors; vola, duravit, d-line, dorma – Fixtures & Fittings; custom pendants by Neri & Hu – Decorative lighting; custom chairs and benches by Neri & Hu – Furniture / 设计时间：2011.9 / 竣工时间：2016.9 / 摄影师：©Pedro Pegenaute

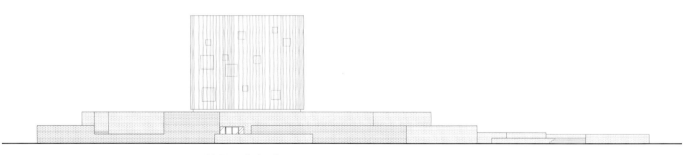

南立面 south elevation

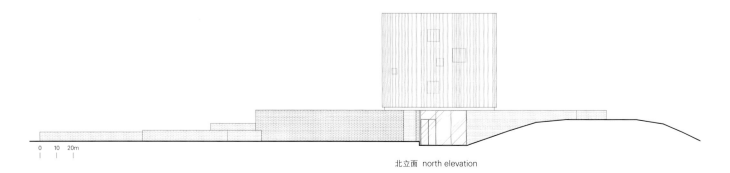

北立面 north elevation

东立面 east elevation

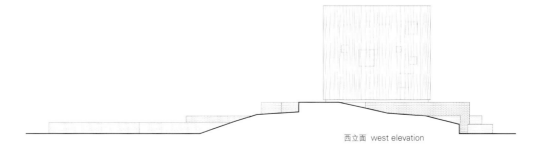

西立面 west elevation

which alternatively hides and reveals. In the daytime, the white box emerges shimmering gently in the sunlight, subtly exposing its contents. At night, the white box becomes a jewel-like beacon in the project, its various windows emitting a soft glow in all directions.

Inside the building, visitors continue on their guided journey through the pre-function area and then into the main chapel space, which features a light-filled 12m high space. There is a seamless integration with the surrounding nature as picture windows frame various man-made and natural landscapes. A mezzanine level hovers overhead to accommodate extra guests, and includes a catwalk encircling the space, allowing 360 degrees of viewing angles. The mezzanine is integrated into a wood louvered cage element which wraps around the whole upper part of the room. A grid of glowing bulb lights and delicate bronze details give a touch of opulence to the otherwise quietly monastic spaces. Custom wood furniture and crafted wood details compliment the simple material palette of gray brick, terrazzo, and concrete.

Another feature of the chapel building is a separate staircase alongside the main space, which allows visitors to gain access to the rooftop for unrivaled views across the scenic lake. Various openings along this stair ascent give unexpected views both internally and externally.

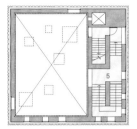

五层 fourth floor

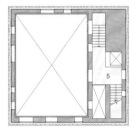

四层 third floor

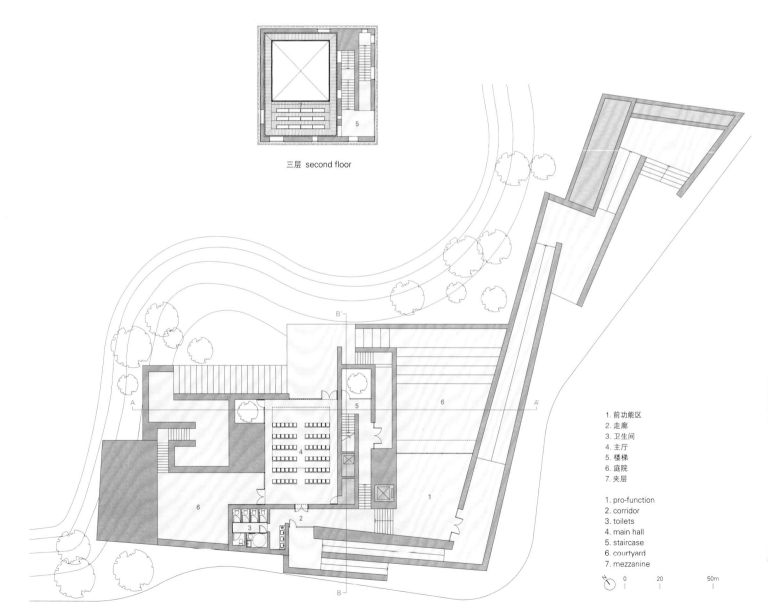

三层 second floor

一层 ground floor

1. 前功能区
2. 走廊
3. 卫生间
4. 主厅
5. 楼梯
6. 庭院
7. 夹层

1. pro-function
2. corridor
3. toilets
4. main hall
5. staircase
6. courtyard
7. mezzanine

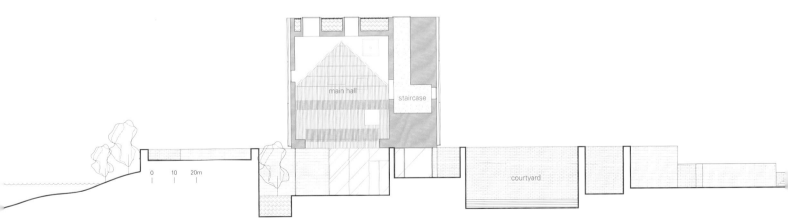

A-A' 剖面图 section A-A'

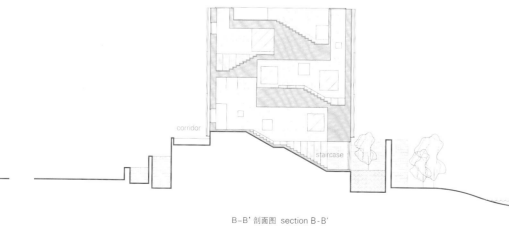

B-B' 剖面图 section B-B'

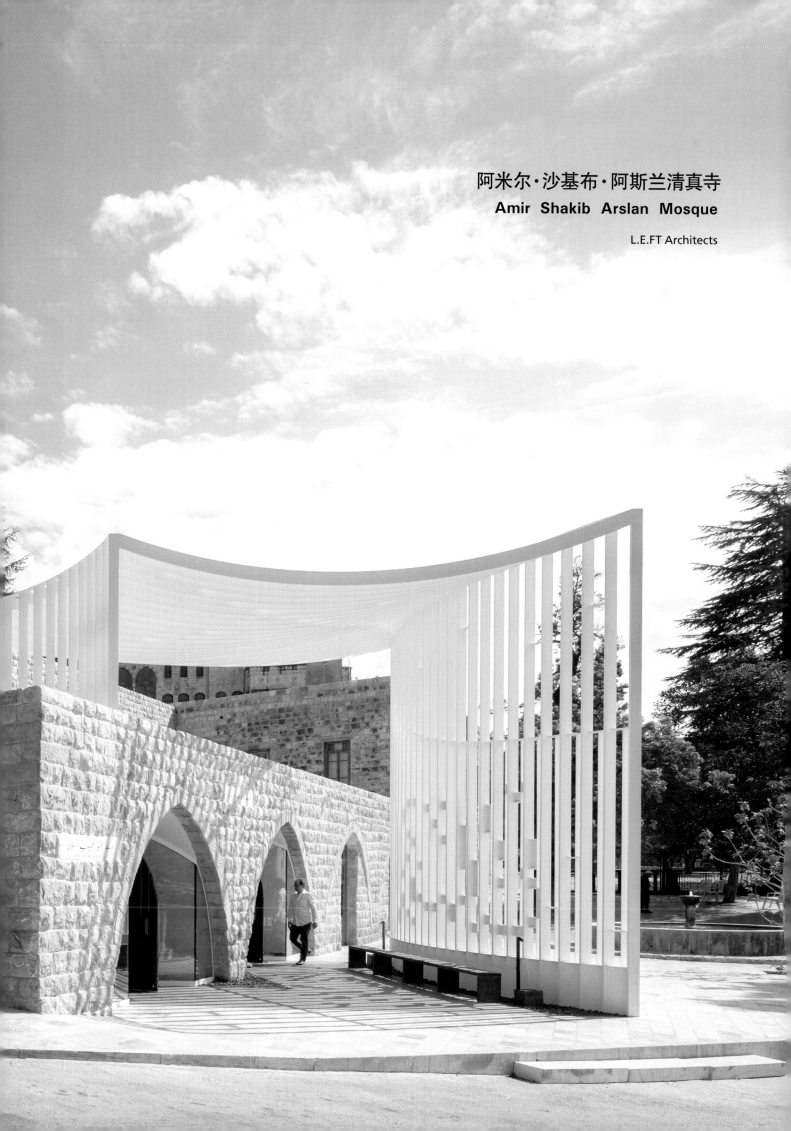

阿米尔·沙基布·阿斯兰清真寺
Amir Shakib Arslan Mosque

L.E.FT Architects

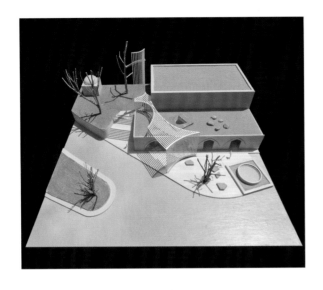

这座只有100m²的小清真寺项目包括对原有的清真寺加以翻新改造，另外在其之上加建一个尖塔，作为一个象征性地标。原有的清真寺是一个带有十字形交叉拱顶空间的砖石结构，紧邻一座18世纪的古老宫殿。翻新改造工作始于拆除20世纪70年代建造的额外楼板，将砖石结构的清真寺恢复到原来的形式。原清真寺前面有一个停车场，现在被建成市民广场，使清真寺的正面变成了一个公共广场，不仅有可供休息的座位，有喷泉，有斋戒沐浴（洗礼）的空间，还可以在新种植的无花果树下纳阴乘凉。

原清真寺上新建的细长尖塔通过一个稍稍呈凹面的遮篷在水平方向上与位于广场层面的一面呈弧线弯曲的墙连为一体，既为清真寺划定出门廊空间，也在清真寺内部和街道之间营造出一个过渡空间，同时增加了清真寺的私密性，使其与外面隔离开来。

清真寺的外围护结构是由一片片白色钢板严丝合缝地组成的，角度与麦加方向平行，朝向麦加的方向。当从一个倾斜的角度看这些钢板时，钢板堆叠在一起，构成一个完整的清真寺体量。从正面看，通过其薄薄的平面，这一清真寺体量消失了，与视觉上饱含历史感的背景融合在一起，让人暂时忘记其真实存在。

这个结构支架将会慢慢爬满常春藤植被，增加建筑与自然环境的融合。薄薄的尖塔和弧形的广场墙上分别装饰着两个词，一个词是自

然,另一个词是人;这个经过折叠和像素化处理的结构支架一是重新诠释了黑格尔的辩证法——自然与人,二是重新唤起长期被遗忘的伊斯兰教信仰的传统——"人道主义"。

至于对原结构内部的处理,建筑师采取最小干预策略,一是使用来自叙利亚阿勒波市的特殊石灰混合物将拱顶的内弧面完全涂成白色,二是在拱顶处开了一个新天窗,使Quiblah墙朝向麦加方向,同时使阳光照射到米哈拉布空间(译者注:Mihrab,米哈拉布,指清真寺正殿纵深处墙正中间指示麦加方向的小拱门或壁龛)。透过天窗,人们还可以看到室外的尖塔,在视觉上将声音和美景连接起来,而在典型的清真寺里,声音和美景往往是分离的。

同样,米哈拉布的凹面弧形墙由反射抛光的不锈钢装饰而成。虽然米哈拉布朝向麦加方向,但在视觉上与更宽阔的背景融为一体,与外部融为一体,使清真寺的内部空间感扭曲,破坏了内部空间的轴对称性。

总的来说,该清真寺的设计是对现代性精神的一种颂扬,因为在结构上,它与抽象的概念、短暂性(无常)的概念有关;从象征意义上来说,它表现了伊斯兰人道主义传统的延续性。它代表了一场思想文化战争的一部分。这场战争需要与不同宗教间的原教旨主义势力进行斗争。在这场战争中,建筑是一种武器。

This small mosque of 100m² included a renovation of an existing masonry cross-vaulted space and the addition of a minaret, grafted onto the existing structure as a symbolic landmark, next to the 18th century old palace. The reconstruction started with the demolition of the 1970's extra floor, restoring the masonry structural to its original form. A new civic plaza was created in what was before an adjoining parking space, turning the frontage of the mosque into a public square with seating, water fountain, ablution space and shading under a newly planted fig tree.

The mosque's new slender minaret is linked horizontally through a gently concave canopy to a curved wall at the plaza level, delineating a portico for the mosque and creating a transitional space between the interior of the mosque and the street as well as adding privacy for the mosque from the outside.

旋转倾斜的清真寺
rotating oblique mosque

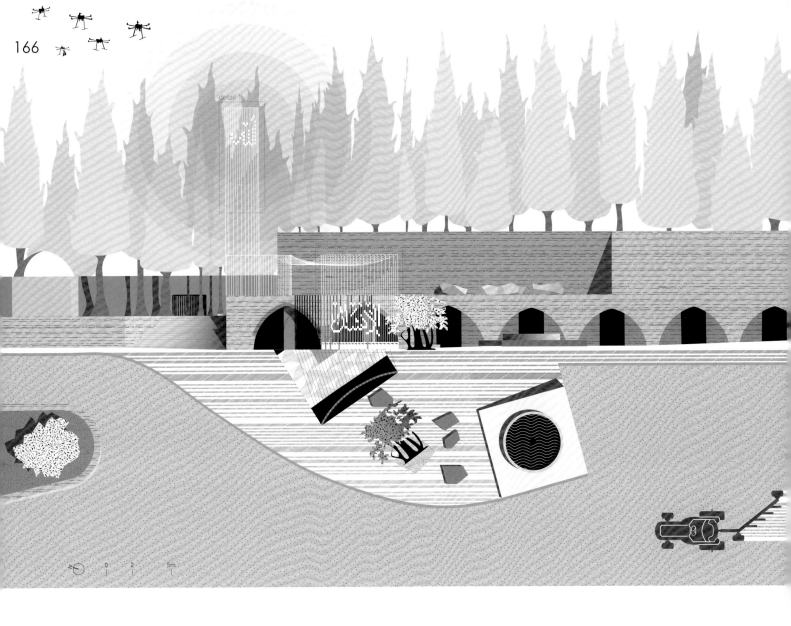

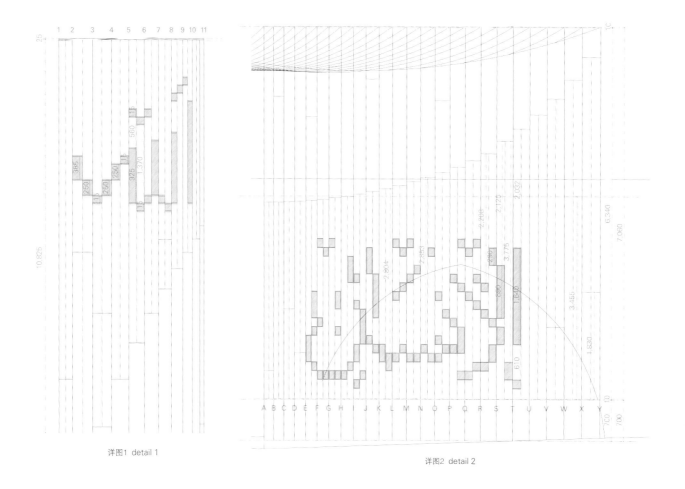

详图1 detail 1　　　　　　　　　　　　　　详图2 detail 2

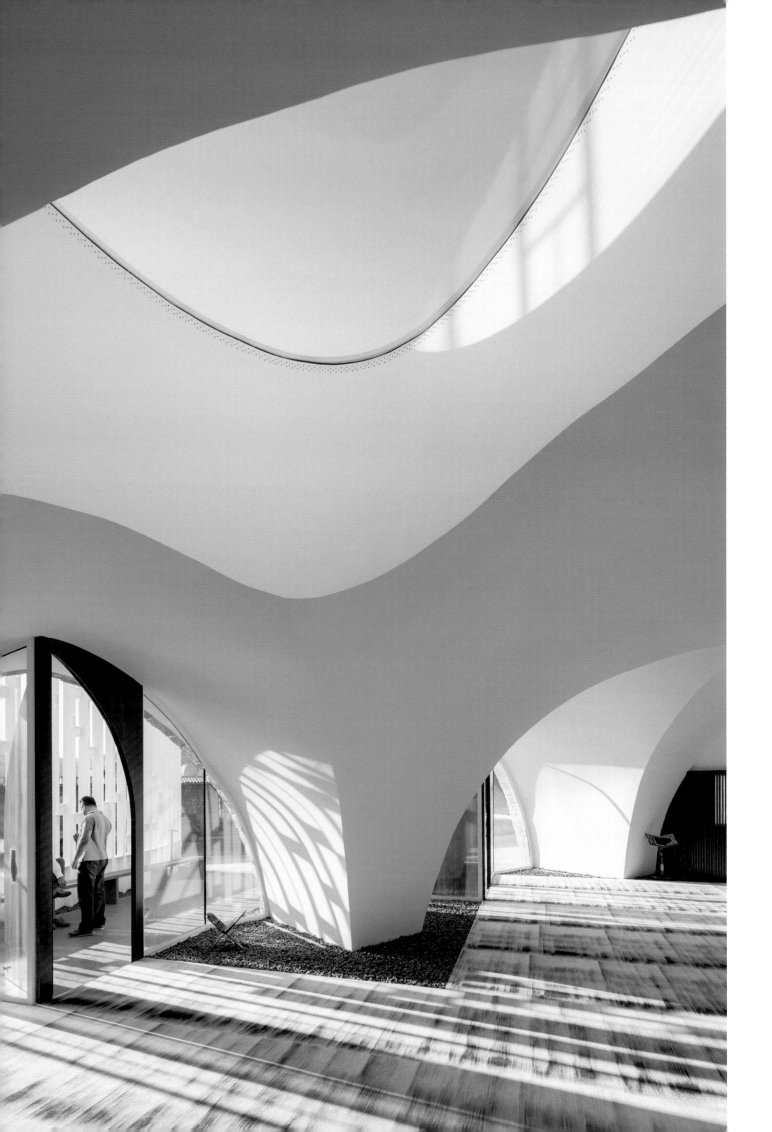

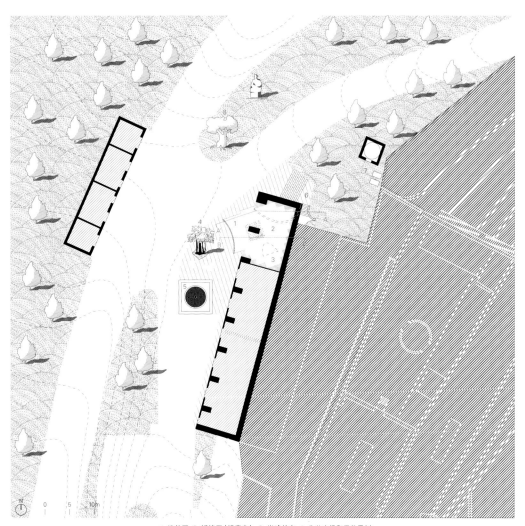

1. 洗礼区 2. 祈祷厅（清真寺） 3. 米哈拉布 4. 公共广场和无花果树
5. 修复后的建于约18世纪的游泳池 6. 通往屋顶花园的楼梯 7. 陵墓花园 8. 橄榄树
1. ablution area 2. prayer hall (Mosque) 3. Mihrab 4. public plaza and fig tree
5. revived 18th c. pool 6. stair to roof garden 7. Mausoleum garden 8. olive tree

The envelope of the mosque is strictly formed of thinly sliced painted white steel plates, faithfully angled in a parallel direction to Mecca. When looked at obliquely from an angle, the steel plates stack to compose a complete and comprehensive volume of the mosque. Looked at frontally, the mosque's volume, through its thin planarity, disappears and blends with its visually rich historical backdrop, momentarily suspending belief in its actual presence.

The structure armature is to be slowly populated by ivy vegetation, increasing its integration with its natural setting. The thinly shaped minaret and the curved plaza wall are structured by integrating the two words Nature and Human Being; bi-folded and pixelated into the steel armature to recreate the Hegelian dialectic of Nature/Man, but also to recall the long forgotten "humanist" tradition of the Islamic faith.

On the inside of the existing structure, the minimal intervention involved a "white-out" of the concave surfaces of the vaults, using special Lime mix brought from Aleppo in Syria, as well as the introduction of a new skylight that cuts the vaulted space to register the direction of the Quiblah wall towards Mecca, and bring light towards the Mihrab space.

Through the skylight, one can see the minaret in a visual looping of exterior back to the interior, linking visually the disassociation in typical mosques between the sound and the vision.

Similarly, the Mihrab is articulated with a concave reflective polished stainless steel arched wall that, though pointing towards Mecca, implodes this axiality by merging it visually with the wider context, bringing outside in, and distorting the interior spatiality of the mosque.

Overall the design of the mosque is a celebration of the ethos of modernity as it relates tectonically to the notion of abstraction, of ephemerality, and representationally to the continuity of the humanism tradition in Islam. It represents a part of a cultural war of ideas that needs to be fought against the fundamentalist forces across religions, a war where architecture is a weapon.

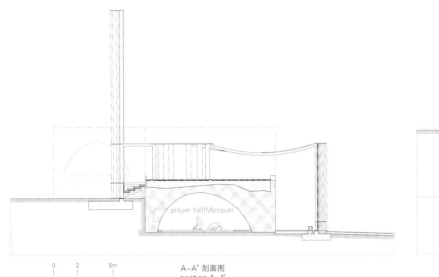

section A-A'

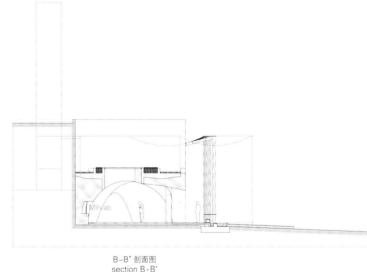

section B-B'

项目名称：Amir Shakib Arslan Mosque / 地点：Moukhtarah, Shouf Mountains, Lebanon / 建筑师：Makram el Kadi, Ziad Jamaleddine-L.E.FT Architects
项目团队：Gentley Smith, Rafah Farhat, Elias Kateb, Alex Palmer, Nayef al Sabhan,Tong Shu, Shun-Ping Liu / 合伙人负责人：Makram el Kadi, Ziad Jamaleddine
保存建筑师：L.E.FT with Zaher Ghosseini / 业主代表：Zaher Ghosseini / 金属制造：ACID-Karim Chaya / 生产商：Karim Chaya-ACID
工程师：Antoine Bou Chedid / 照明设计：Maurice Assoc-Hilights / 景观建筑师：L.E.FT Architects / 总承包商：ACON / 地毯设计：L.E.FT with Lawrence Abu Hamdan / 地毯制造：Moooi / 祷告人员：Lawrence Abu Hamdan and Nisrine Khodr / 面积：100m² / 客户：Walid Jumblat / 施工时间：2015.3—2016.9
摄影师：©Iwan Baan (courtesy of the architect)-p.160~161, p.162, p.166, p.168, p.169right-lower, p.170; ©Ieva Saudargaite (courtesy of the architect)-p.164~165, p.167, p.169right-upper

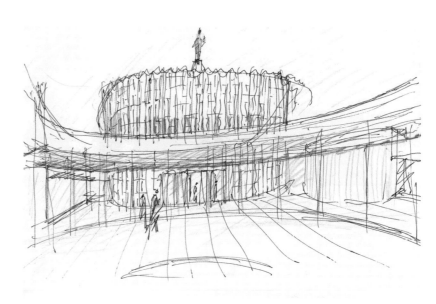

马里博尔市的唐博斯克教堂
Don Bosco Church in Maribor

Dans Arhitekti

唐博斯克教堂建在马里博尔市的居民区中，恰好坐落于波霍列山的山脚下。这是新教区中心建设的一部分，更是欣赏派克斯卡·戈尔卡山美景的绝佳观赏点。

这一属于慈幼会教徒社区的宗教综合体的平面设计相对封闭，好似一个独立的岛屿，四周环绕着一个小公园，这样可以避免附近两条道路的干扰。该建筑本身是一个矩形建筑体块，结构清晰，内部有一个封闭的中央庭院，中殿和钟楼高耸其上。

庭院中栽有一棵老菩提树，人们可以在此静思冥想。穿过庭院，教堂就豁然出现在柱廊后。只需一眼，你就可以感受到教堂中殿这一空间的延展性。中殿从地面开始向上延伸，越过整个矩形综合体低层建筑的屋顶，一直延伸到王冠状的屋檐。中殿体量嵌置在低层建筑的结构中。其立面全部由砖镶嵌，装饰有七个由釉面砖制成的十字架。

中殿的平面形状是一个圆角圆边的三角形，是一个反省自我、静思冥想的地方，因此唯一可见的视野是上方的天空。这种明亮强烈的光线营造出了整座教堂最为神圣的地方。

建筑师使用了多种不同的光效，来突出中殿怀抱式的形状。他们精心设计了一处灯光秀：圆形天窗中透过生动的自然光，与中殿的漫射光交织在一起，司祭席和后面唱诗班散发出更加柔和的自然光。这一设计看似简单，实则暗含匠心，让建筑打破了自身的限制，成为一个装满优美光线的容器。

中殿的结构设计简洁质朴，毫无雕饰，最突出的特点就是强大的可塑性，这是由形状和缓的混凝土塑造出来的。在中殿的椭圆形边缘上，建筑师在水平方向上开了几个凹槽，有的用于摆放神龛，有的是圣器室和忏悔室的入口，有的是教堂两侧唱诗班以及教堂后面唱诗班的入口，还有一个是做圣餐祈祷的地方。这些凹槽开口给人一种围绕中央空间平静旋转的感觉。教堂的木长椅可以坐300人。地板采用的是厚实的橡木地板，上方悬挂着100盏纯手工制作的陶土灯。橡木地板和陶土灯这样暖色系的材料为教会社区营造出一种充满安全感的氛围。建筑师还设计了所有用于礼拜仪式的设备，如圣坛、读经台、神龛、牧师席、长凳，还有艺术品的摆放。

神龛放置在礼拜堂的侧面，看起来就是一幅三联画。两个柜门遮挡着中间的柜子，打开两个柜门，可以看到里面神秘的光源——永恒之光。里面的神龛由一个珍贵的圣像装饰着——都灵圣母进教之佑教堂之前使用的神龛门。圣母进教之佑教堂由约翰·博斯克设计建造。

The Don Bosco Church is part of a new parish center, located in a residential quarter of Maribor, under the Pohorje Massif. The site boasts a view towards the Pekrska Gorca Mountain.

The religious complex belonging to the Salesian community is designed as a closed building island, placed inside a small park, which protects the complex of two nearby roads. The building itself is a clearly articulated rectangular architectural mass with an introverted central courtyard with the nave and the bell tower rising above it.

Upon entering the contemplative courtyard adorned by an old linden tree, the church suddenly becomes visible across the portico. At a glance, one can feel the extent of the nave from the ground over the roof of the lower part of the complex all the way to the cornice that represents a crown. The volume of the nave is set into the structure of the lower one-storey building. Its uniform brick facade is embellished with ornaments representing seven crosses made of glazed bricks.

The shape of the nave is derived from a triangle with rounded corners and sides. The nave is the place for introverted contemplation, therefore the only view out is towards the sky. The most sacred place is thus marked by an intense

项目名称：Don Bosco Church in Maribor
地点：Engelsova ulica 66, Maribor
建筑师：DANS arhitekti
项目管理：Rok Bogataj, Miha Dešman, Eva Fišer Berlot, Vlatka Ljubanović, Katarina Pirkmajer Dešman
景观建筑：Mojca Balant
静态学设计：Franc Žugel
电力工程师：Zoran Pavlin, Jože Velnar
机械工程师：Rudi Pavlić, Aleš Matuš
建筑物理：Peter Žargi
客户：The Salesians in Slovenia
用地面积：14,067m²
建筑面积：ground floor 615m²; attic 490m²
结构：Concrete
材料：concrete, oak wood for the floor and for all other elements of interior (doors, benches, etc), floor and all elements of prezbiterium in terazzo
造价：2,000,000 EUR
竞赛时间：2007 / 项目时间：2011
竣工时间：2015
摄影师：©Miran Kambić (courtesy of the architect)

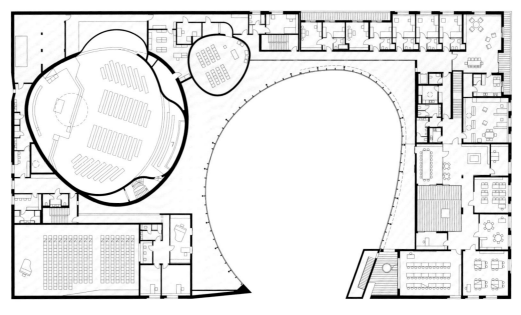

二层 first floor

1. 主入口	1. main entrance
2. 侧入口	2. side entrance
3. 圣坛	3. chancel
4. 神龛	4. tabernacle
5. 侧面唱诗班	5. side-choir
6. 洗礼室	6. baptistery
7. 后面唱诗班	7. back-choir

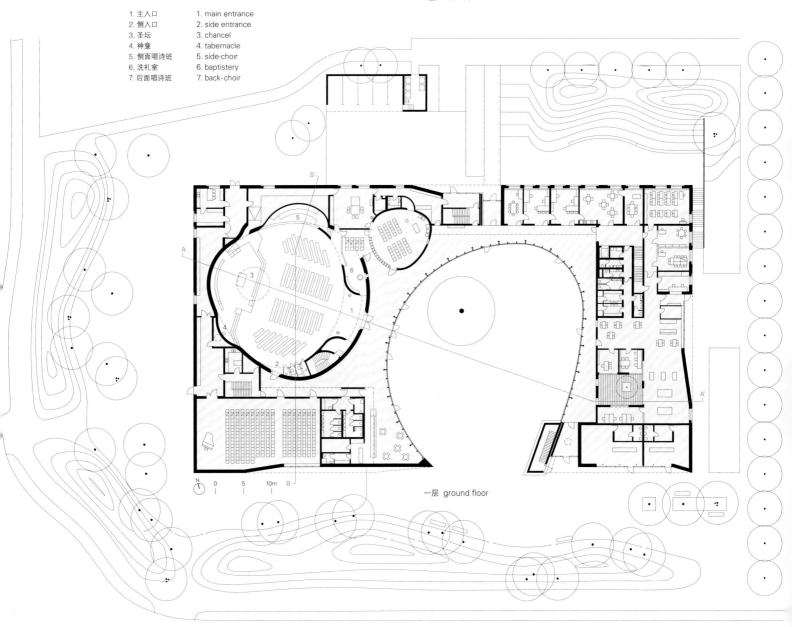

一层 ground floor

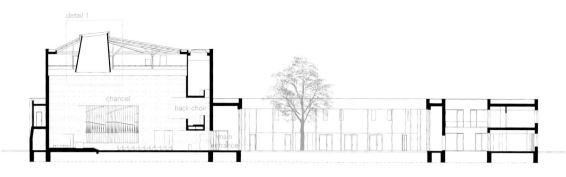

A-A' 剖面图 section A-A'

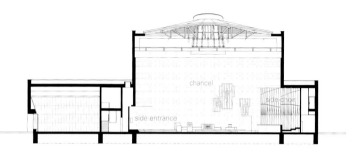

B-B' 剖面图 section B-B'

experience of light.

The architects used different qualities of light in order to accentuate the hugging shape of the nave. They choreographed a play of light: vivid natural light emitting from the round skylight interferes with diffused light of the nave. More soft natural light is coming from behind the presbytery and the rear choir. This simple yet so complex design allows the building to transcend its physical presence, becoming a mere container of light.

Pristine non-decorated expression of the structure of the nave is accentuated by the plasticity of softly shaped concrete. The oval perimeter of the nave features horizontal lateral recesses, which house the chapel with the tabernacle, entrance to the sacristy, confessionals, side and rear choirs and a place for communion prayers. These openings induce a feeling of peaceful rotation around the central oculus. The church seats 300 people on wooden benches. The floor below is covered massive oak wood floor while 100 handmade clay lamps from the horizon above. Warm materials such as oak wood fittings and clay lamps create a feeling of safe haven for the church community. The architects also designed all the liturgical equipment, such as the altar, the ambo, the tabernacle, sedilia, benches and layout of the artworks.

Tabernacle, placed in the side chapel, is a triptych with the outer wings veiling the central cabinet. The tabernacle opens in two sequential steps: when the wings are closed, they reveal a floating gold rectangle. Upon opening the two side wings, the interior with the mysterious light source – the eternal light – is revealed. The inner tabernacle is adorned with a precious icon – former tabernacle doors from the Basilica of Our Lady Help of Christians, in Turin, which was built by John Bosco.

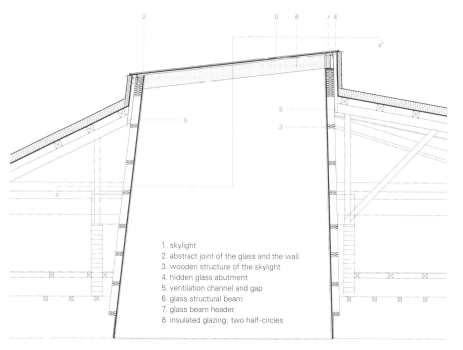

1. skylight
2. abstract joint of the glass and the wall
3. wooden structure of the skylight
4. hidden glass abutment
5. ventilation channel and gap
6. glass structural beam
7. glass beam header
8. insulated glazing: two half-circles

详图1 detail 1

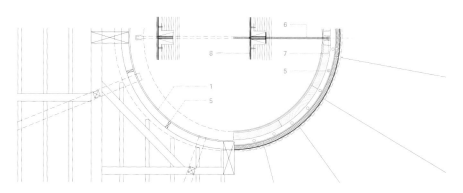

a-a' 详图 detail a - a'

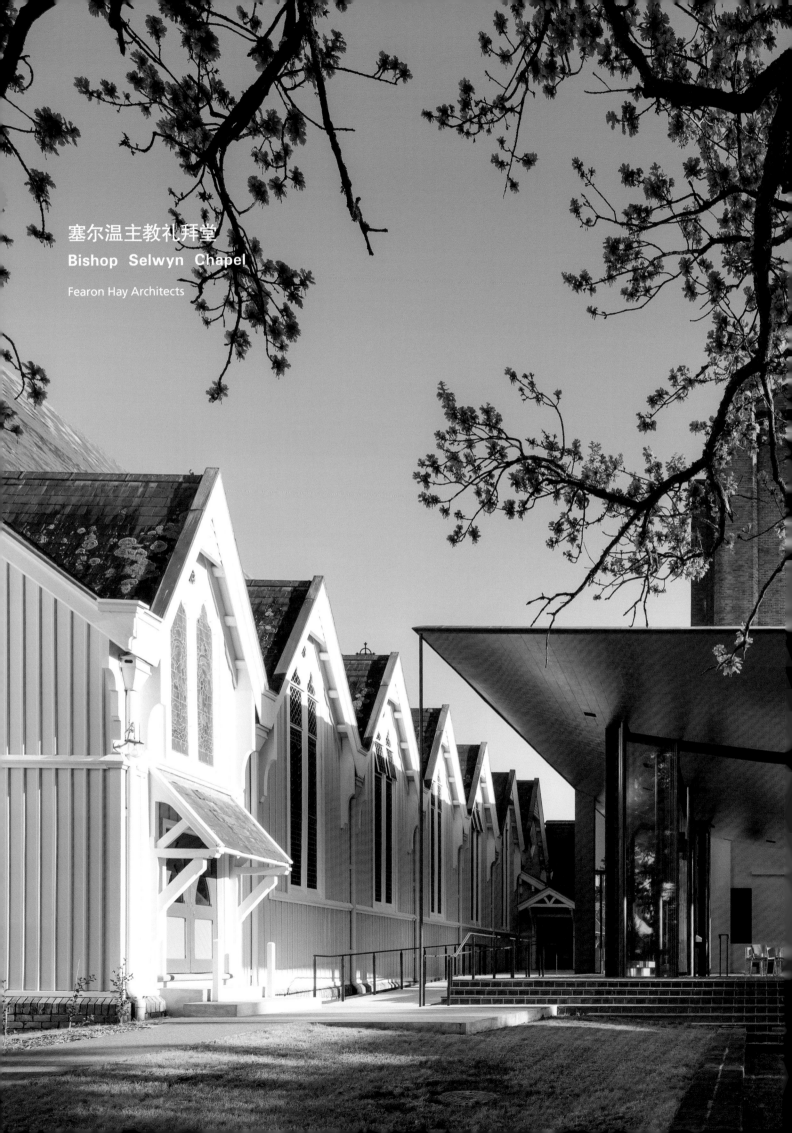

塞尔温主教礼拜堂
Bishop Selwyn Chapel

Fearon Hay Architects

1. 圣三一大教堂 2. 塞尔温主教礼拜堂 3. 圣玛丽教堂
1. Holy Trinity Cathedral 2. Bishop Selwyn Chapel 3. Saint Mary's Church

　　数年前，如果你漫步走在奥克兰大教堂的南面周边，就可以欣赏到有着数百年历史的一棵棵橡树、美丽精致的纪念花园、历史悠久的木质小教堂以及一座尚未完工的大教堂。

　　这些空间朝向南面，照不到阳光，远离街角的正面及公共场所的入口。这里一片空旷，尚未与由石灰石建造的拱形主圣坛连为一体，四周被围着。但是，这个地方很漂亮，透过橡树林，可以远眺到奥克兰著名的火山地标，芒格基基正好位于其南面的轴线上。

　　设计竞赛的要求是要在奥克兰市中心一个重要的地块上完成圣三一大教堂的建造，同时在这儿新建一座小礼拜堂，为该场地增添更多的建筑形式，从而完成这座年轻城市最初1888年修建这座大教堂的空间规划。

　　教区的基本要求是为100人提供一个做礼拜、唱赞美诗、表演和举行活动的空间。而建筑师提供的远非如此，他们设计了一个外部平台、门廊和楼梯台阶，通向低处的草坪，与放置骨灰的庭院里低矮的玄武岩墙连为一体。

　　小礼拜堂的布局与圣坛相呼应，并与圣坛成为一体。延伸的回廊使身在小礼拜堂的人们仍可以欣赏到美丽的橡树和花园的景色。在小礼拜堂的两侧，新铺设的地面沿着台阶一直延伸至花园中。

　　由艺术家尼尔·道森设计雕刻的十字架矗立在小礼拜堂的外面，掩映在百年老橡树的树冠中，使小礼拜堂与庭院成为一体。小礼拜堂小巧简约，四个立面全是玻璃，使其内部与回廊隔离开来。小礼拜堂四面透明，但很有层次感：有圣玛丽教堂的维多利亚式屋顶轮廓，有精细的木工制作，透过小教堂玻璃窗还可以看到随季节变化而变化的橡树树冠。小礼拜堂南立面的透明玻璃使其与外面环境融为一体，封闭性完全消融：玻璃门向两边滑开，就消除了礼拜堂内部与外部花园和远处的火山景色之间的唯一屏障。

　　宽大的金色雨篷从原有圣坛的砖墙位置向前方延伸，通向花园，覆盖了一个富有开放性的空间，在这里可以看到圣玛丽教堂哥特式的木质装饰，可以看到公园和如雕塑般存在的橡树，还可以看到奥克兰的火山口和海港的优美风光。这个金色雨篷重整了空间序列，邀请花园里的人来到这里，花园也吸引雨篷下的人走进花园，它还与历史建筑圣玛丽教堂结合在了一起。

　　该设计旨在吸引人们进入小礼拜堂，同时它还与外面的自然美景融为一体，实现了"在花院中做礼拜"这一原始设计概念。

Walking the southern perimeter of Auckland's Cathedral grounds years ago would be an experience of the eclectic arrangement of 100 year old oaks, a beautifully subtle memorial garden, a historic wooden church and an unfinished Cathedral.

The spaces are to the south, out of the sun and away from the street corner frontages and public entrances. They are unoccupied, the unfinished connections to the interior of the main chancel limestone vaulting boarded up. But the site is beautiful – the oaks frame an elevated view of the volcanic landmarks of Auckland with the iconic Maungakiekie directly on its southern axis.

The competition to complete the Holy Trinity Cathedral on a prominent ridgeline in central Auckland sought to construct a new chapel, adding further built form to the site and in doing so completing the spatial programme of the original 1888 vision for a Cathedral in the young city.

The diocese' brief required a worship, choral, performance and event space for 100. The approach extends this to further provide an external terrace, covered porch and stairs that descend into the grassed spaces and low basalt walls of the columbarium garden.

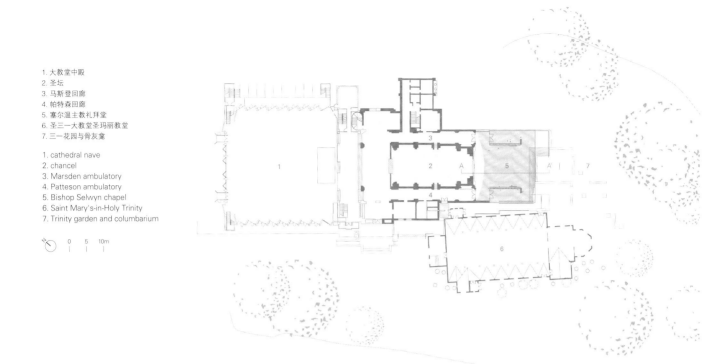

1. 大教堂中殿
2. 圣坛
3. 马斯登回廊
4. 帕特森回廊
5. 塞尔温主教礼拜堂
6. 圣三一大教堂圣玛丽教堂
7. 三一花园与骨灰龛

1. cathedral nave
2. chancel
3. Marsden ambulatory
4. Patteson ambulatory
5. Bishop Selwyn chapel
6. Saint Mary's-in-Holy Trinity
7. Trinity garden and columbarium

圣三一大教堂圣玛丽教堂
Saint Mary's-in-Holy Trinity

塞尔温主教礼拜堂
Bishop Selwyn Chapel

详图1 detail 1

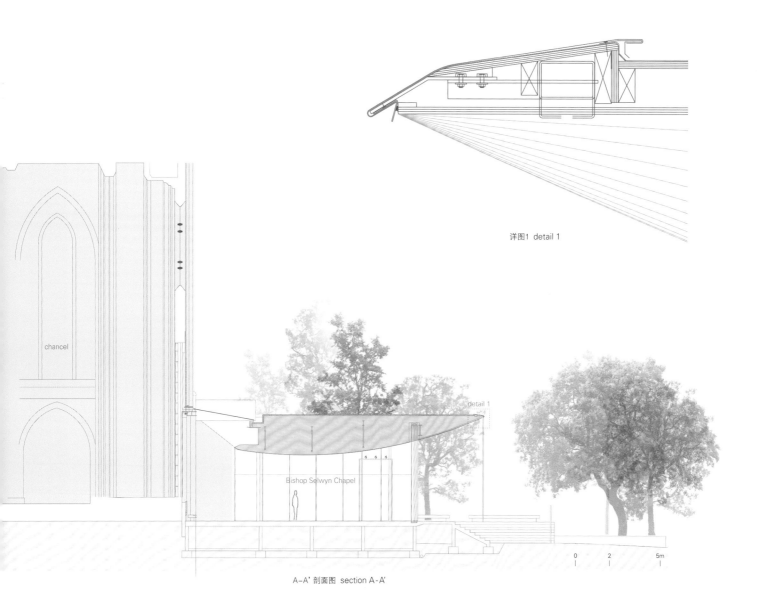

A-A' 剖面图 section A-A'

The chapel is laid out in response to the Chancel to which it is joined. The extension of the ambulatories now provides sightlines to the oaks and garden from deep within this interior and the new floor plane continues either side of the chapel as terraces flowing into the garden.

A cross rendered in a sculptural work by artist Neil Dawson is positioned beyond the internal enclosure, anchoring the space between the built form of the Cathedral and the canopy of oaks and emphasizing the garden connection. The enclosure itself is minimal – glazed planes flank the space and define the chapel from its ambulatories. This enclosure is dematerialised yet layered: Saint Mary's Victorian roof profile and timber detailing combine with the seasonally changing canopies of the oaks reflected on and seen through the transparency of the chapel. The southern enclosure is also glazed, but in this case the containment can be dissolved, the glass panels sliding away to either side to remove any physical barrier to the garden and the volcanic landscape of Auckland beyond.

The sequence of space, invitation to and from the garden, and inclusion of historic Saint Mary's are brought together beneath a broad golden canopy, extended as a simple plane from the vertical brick mass of the existing chancel and draped into a form that opens towards its edges – opens to include the gothic timber profile of Saint Mary, opens to the garden and sculptural presence of the oaks and opens to the Auckland landscape of volcanic cones and harbour.

The result seeks to be a presence that is both an invitation to the space within and a place from which to connect to the natural beauty beyond the building – fulfilling the original concept of the brief as "worship in the garden".

项目名称：Bishop Selwyn Chapel / 地点：446 Parnell Road, Parnell, Auckland / 建筑师：Fearon Hay Architects / 项目团队：Jeff Fearon, Tim Hay, Stephen de Vrij, Michael Huh, Chris Wood / 项目管理：MPM Projects / 结构工程师：Holmes Consulting Group / 机械工程师：Mott MacDonald 电力工程师：Mott MacDonald / 照明工程师：Lightworks and ECC Lighting / 景观建筑师：Jacky Bowring / 客户：Holy Trinity Cathedral / 用途：Community 总楼面面积：370m² / 室外饰面：Precast concrete sandblasted / 室内饰面：Plaster and faux stone plaster / 材料：Gold leaf, Brass, Stone tiles, Pre-cast concrete, Faux stone plaster, Plaster, Wall louvers / 造价：Withheld / 竣工时间：2016
摄影师：©Patrick Reynolds (courtesy of the architect)

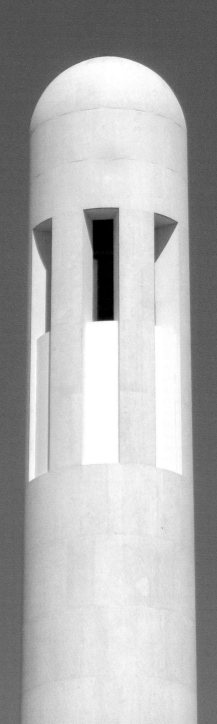

姆什莱布清真寺
Msheireb Mosque

John McAslan + Partners

新建的姆什莱布清真寺坐落于姆什莱布遗产区域内。在当地居民的心目中，这一区域一直都是宗教和政治力量的有力支柱。而新建的姆什莱布清真寺的建筑设计特别采用了卡塔尔独特的建筑材料和建筑细节处理，将这种历史传承下来的建筑体量和空间布局与现代主义融合在一起。

设计方法

几百年来，传统的卡塔尔清真寺一直都运用方向控制、遮阳、自然通风和水环境为前来祷告的信徒营造一种宁静祥和的祈祷氛围。而新建的姆什莱布清真寺不论是建筑风格、整体布局，还是大厅装饰，都继承了卡塔尔清真寺这种优良的传统。此外，新建的姆什莱布清真寺设计也反映了伊斯兰艺术和建筑文化的几项重要原则：造型简洁、功能完整、精神引导、光线营造、图案华丽、几何叠加以及水环境的穿插。新建的姆什莱布清真寺的平面布局以双矩形为基础，与几何图案和花纹设计的手法一样，都遵循了伊斯兰清真寺的经典设计先例，创造了一个拥有完美比例关系的优雅空间，光线透过叠加不同花纹的屋顶形成了一幅光影斑驳的美好画面。

建筑材料和建筑方法

新建的姆什莱布清真寺是由纯净的白色石材构成的，就像是一个清爽纯净、完美无缺的立方体。清真寺的入口和寺内院落四周都安装有刻着伊斯兰图案的金属大门。走进祈祷大厅，太阳光透过绘有图案的镂空屋顶，在大厅的地面上形成一幅光影交织的图画，营造出一种半静而美好的氛围，为祈祷提供充足的沉思空间。石柱排列在庭院两侧，围合成一个堪称完美的庭院广场。祈祷大厅入口前的浅水池也为庭院广场增添了些许平静沉思的感觉。石质的尖塔截面是一个圆形，越往塔尖，塔身越细，这就要求每块石头都要被切割成不同的形状，这样才可以使塔身的半径随着塔身高度的增加而不断减小，从而构造出石塔独特的整体外观。

可持续性设计

新建的姆什莱布清真寺采用现浇混凝土作为框架，然后再用砌块进行墙体填充。接下来，采用当地的石灰石作为建筑墙体外层的覆层，选用卡塔尔特制的石材铺设庭院地面。幕墙用铸铜制成，看起来显得华美而深邃。新建的姆什莱布清真寺是按照LEED金奖标准设计的，采用被动式和主动式相结合的可持续建造技术。太阳能光伏和太阳能热水器的应用就很好地体现了这两种设计方法的结合使用。祈祷大厅的设计也采用了被动式照明设计，白天无须进行人工照明。这座新建筑的整体外观和内部配置实际上都是以传统的卡塔尔清真寺作为基本标准的。几百年来，传统卡塔尔清真寺一直都通过采光、遮阳、自然通风和水环境为前来祷告的信徒营造舒适的礼拜环境。

The new Msheireb Mosque is located within the Msheireb Heritage Quarter. The area has always been the anchor of religious and political power for the local population. The design fuses Modernism with a historical arrangement of volumes and spaces, using specifically Qatari materials and architectural details.

Design Approach

The style, layout, and decoration of the building is based on traditional Qatar mosques, which have for centuries used orientation, shading, natural ventilation and water to create environments for prayer. The design reflects the key principles of Islamic art and architecture – simplicity, functionality, spirituality, light, pattern, geometry and water.

项目名称：Msheireb Mosque / 地点：Doha, Qatar / 建筑师：John McAslan + Partners / 执行建筑师：Burns & McDonnell
项目管理：TIME Qatar, Turner Construction / 当地顾问：Arab Engineering Bureau / 结构工程师：CE Anderson Associates
照明工程师：GIA-Equation / 景观建筑师：John McAslan + Partners / 承包商：QACC / 当地顾问：Arab Engineering Bureau
客户：Msheireb Properties / 用地面积：4,605m² / 总楼面面积：1,400m² / 材料：Omani limestone, cast bronze doors and black granite water basin, polished plaster internal walls / 项目时间：2010—2016 / 摄影师：©Hufton and Crow (courtesy of the architect)

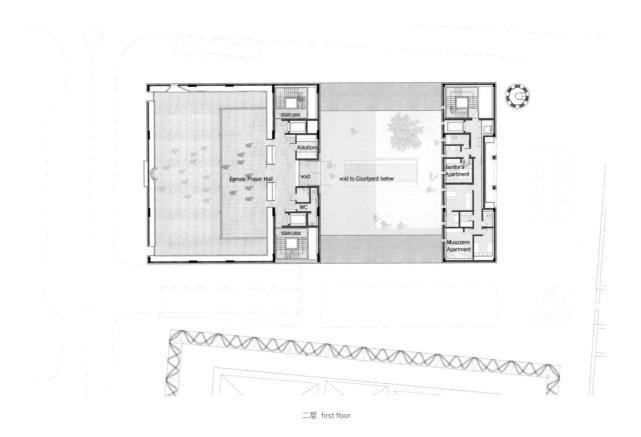

二层 first floor

一层 ground floor

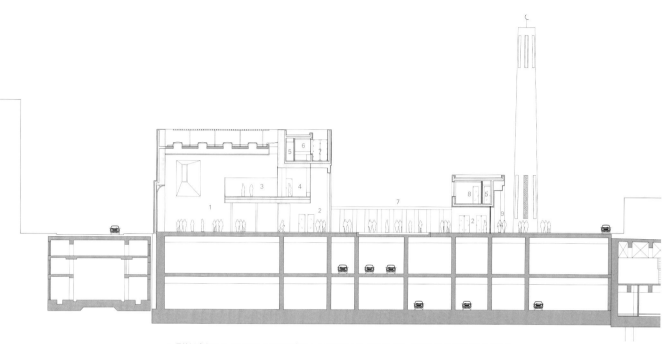

1. 男性祈祷大厅 2. 入口大厅 3. 女性祈祷大厅 4. 女性门厅 5. 门厅 6. 厨房 7. 庭院 8. 门卫起居室 9. 主入口
1. male prayer hall 2. entrance lobby 3. female prayer hall 4. women's foyer 5. lobby 6. kitchen 7. courtyard 8. janitor's living 9. main entrance
A-A' 剖面图 section A-A'

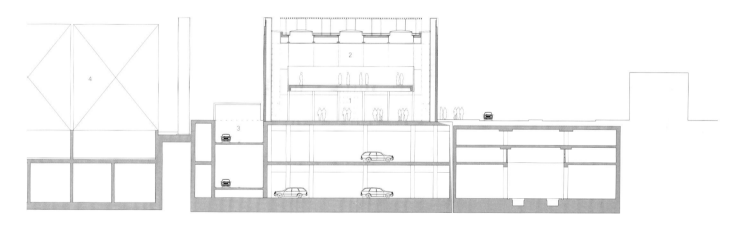

1. 男性祈祷大厅 2. 女性祈祷大厅 3. 汽车坡道 4. 冷却塔
1. male prayer hall 2. female prayer hall 3. car ramp 4. cooling tower
B-B' 剖面图 section B-B'

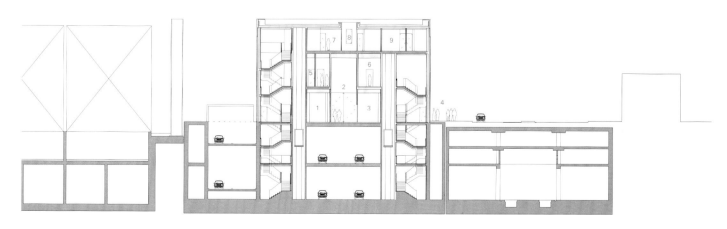

1. 存鞋处 2. 入口大厅 3. 图书馆 4. 女性入口 5. 门厅 6. 洗礼室 7. 伊玛目起居室 8. 庭院 9. 卧室
1. shoe store 2. entrance lobby 3. library 4. female entrance 5. lobby 6. ablution 7. Imam's living 8. courtyard 9. bedroom
C-C' 剖面图 section C-C'

The plan form, based on a double square, follows classical Islamic precedent, as does the use of geometric patterns and designs, creating an elegant space with perfect proportions, with pierced roof screens creating patterns of dappled light and shade.

Materials and methods of construction

The perfect cube building is constructed of crisp white stone. Metal Islamic patterned gates enclose the entrance pavilion and courtyard. Within the prayer hall a perforated, patterned roof allows dappled natural light to illuminate the prayer hall, providing a contemplative space for prayer. A colonnade of stone wraps the courtyard on both sides, framing a perfect courtyard square. A pond and rhyll create a sense of calm and contemplation before the entrance to the Prayer Hall. The stone minaret is circular in section and tapers towards the top, requiring each course of stones to be cut differently to achieve the overall form due to its reducing radius.

Sustainable Design

The Mosque has been constructed using an in situ concrete frame with blockwork infill. Regional limestone is used as cladding and Qatari stone used as accent banding to the courtyard floor. Screens are cast bronze to create richness and depth. The Mosque has been designed to "LEED" gold standard and utilises passive and active sustainable techniques including photovoltaics and solar hot water heaters. The prayer hall is designed so that no artificial lighting is needed during daylight hours. The form and configuration of the building is based on that of traditional Qatari mosques, which have for centuries used orientation, shading, natural ventilation and water to create comfortable environments for prayer.

莱比锡大学帕琳奈教堂
Paulinum, Leipzig University

Design Erick van Egeraat

　　由荷兰建筑师Erick van Egeraat设计的新主楼、大教室和帕琳奈教堂的建成标志着修缮一新的莱比锡大学顺利完工。莱比锡大学始建于1409年，是德国第二古老的高等学府。

　　新建成的帕琳奈教堂地处前宝莱纳教堂的遗址上。尽管宝莱纳教堂是第二次世界大战期间唯一完好无损的教堂，但是在1968年还是被拆除。Erick van Egeraat设计的帕琳奈教堂的竣工，为重建莱比锡大学宝莱纳教堂长达数十年的争辩画上了句号。

　　建筑师Erick van Egeraat从没想过要一砖一瓦地恢复宝莱纳教堂原貌，Erick所做的是用更加具有冲击力的建筑唤醒人们对于宝莱纳教堂的记忆。

　　这所位于城市中的大学的改造包括三个主要部分：礼堂（帕琳奈教堂）、主楼和大教室。主楼在几年前便已完工并向学生开放，大教室的内部在2017年也正式向公众开放。而Andachtsraum以其当代设计理念向人们重新定义了古老的大学教堂应该有的样子。像原来的宝莱纳教堂一样，帕琳奈教堂是一个多功能的空间，既可以为师生提供礼拜空间，也可以用来开展学术性活动，举办音乐会以及举行科学性会议。

　　新教堂的拱顶结构以原教堂空间及其交叉拱顶作为设计参考，由白色石膏结构与玻璃圆柱共同支撑。

　　除此之外，礼堂（帕琳奈教堂）由两个完整的部分组成，有世界上最高的透明玻璃拉门。这些门高度达15.5m，将整个教堂空间一分为二：一部分不涉及宗教的"世俗"礼堂区域用来举行各种各样的活动；而另一部分，即所谓的Andachtsraum区域则更适合冥想与沉思。在这里，从被毁教堂的废墟中抢救出来的墓志铭经过人们的精心恢复，被重新展示出来。

　　整个拱顶空间由玻璃圆柱支撑，令人赞叹不已。为了增强整个空间

礼堂门厅，20世纪10年代
foyer of the auditorium, 1910s

卡尔·马克思广场上的大学教堂和主楼，20世纪60年代
university church and main building on the Karl-Marx Square, 1960s

卡尔·马克思广场上的大学塔楼和主楼，20世纪70年代
university tower and main building on the Karl-Marx Square, 1970s

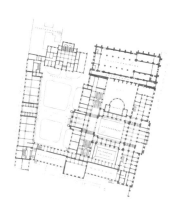

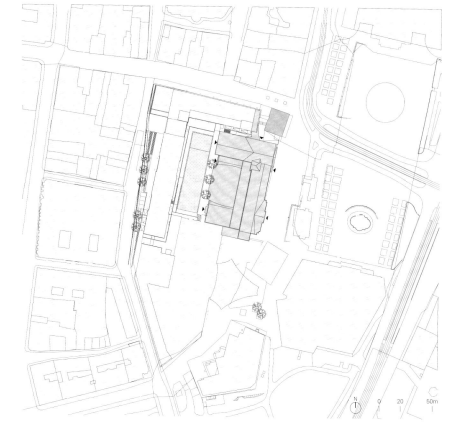

的功能性，有几对玻璃圆柱并没有从天花板一直延伸到地面。无论有无巨大的玻璃拉门的存在，壮观的拱顶空间都能使人们感受到教堂的整体设计感。

　　随着该项目的最后一部分——帕琳奈教堂和Andachtsraum——正式向公众开放，位于莱比锡奥古斯都广场的新莱比锡大学重获其尊贵的身份，并且名正言顺地向莱比锡、向德国乃至向世界宣告其重要地位。

With the new Main Building, Audimax and Paulinum, designed by the Dutch architect Erick van Egeraat, the new University of Leipzig, founded in 1409 and the second oldest university in Germany, has completed.

The new Paulinum of University of Leipzig is situated on the site of the former Pauliner Church, the only church to remain undamaged during the second world war. In 1968 the complex was willfully demolished. With the design made by Erick van Egeraat, decades of debate about the reconstruction of university's Pauliner Church came to an end.
Erick van Egeraat did not propose to rebuild the church stone by stone, but instead created a new building powerful enough to revive the memories of what once was.
The inner-city university redevelopment consists of three main elements: the Auditorium (Paulinum), the Main Building (Augusteum) and the Audimax. While the Main Building

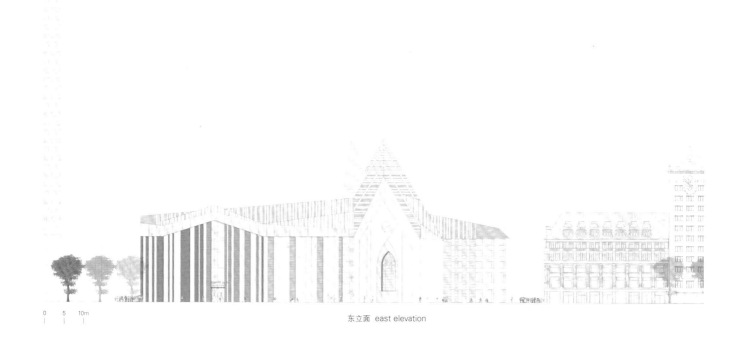

东立面 east elevation

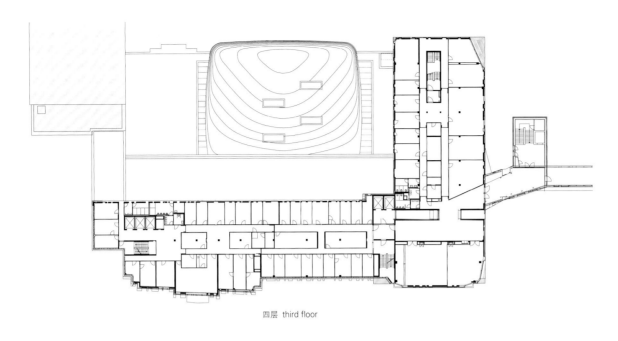

四层 third floor

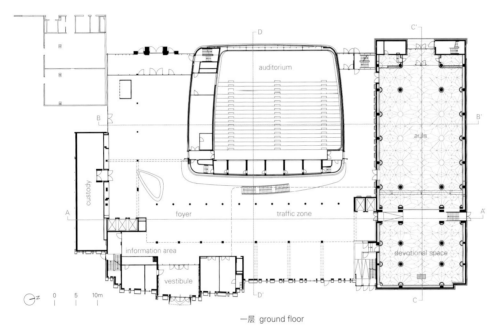

一层 ground floor

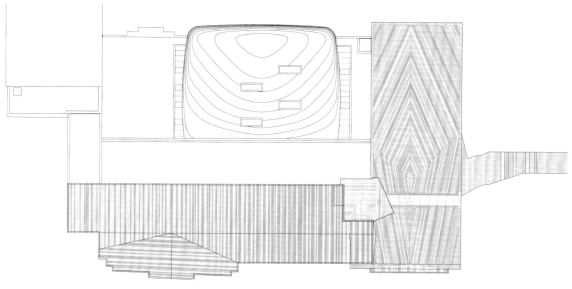

屋顶 roof

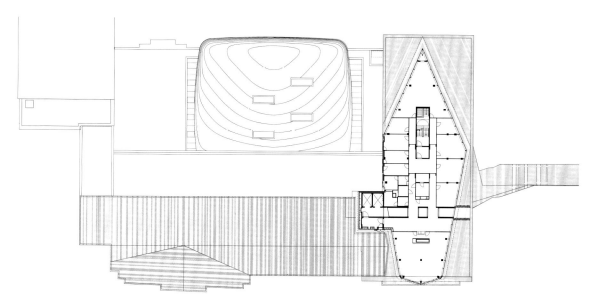

五层 fourth floor

项目名称：Main building and Auditorium of University of Leipzig / 地点：Leipzig, Germany / 建筑师：Design Erick van Egeraat
项目团队：Ekkehard Krainer-Project architect, Harry Kurzhals, Jakub Seiner, Massimo Bertolano, Rene Dlesk, Igor Hobza, Helena Simova, Michael Franke, Claus Müller, Holger Achterholt / 项目管理：Design Erick van Egeraat / 结构工程师：Statikbüro Lochas Forner / 生产商：MLT Ingenieure / 机械工程师：MLT Ingenieure / 电力工程师：MLT Ingenieure, Klett Ingenieure / 照明工程师：MLT Ingenieure / 景观建筑师：Behet Bondzio Lin Architekten / 客户：Sächsisches Staatsministerium der Finanzen vertreten durch Staatsbetrieb Sächsisches Immobilien-und Baumanagement / 用途：university / 总容积率：76% / 建筑规模：one story below ground, four stories above ground / 结构：reinforced concrete
室外饰面：stone, glass, metal, PMMA / 室内饰面：plaster, stone, wood, glass / 竞赛时间：2004 / 竣工时间：2017
摄影师：©J Collingridge (courtesy of University Leipzig) (except as noted)

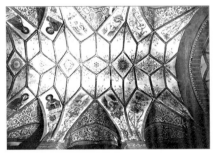
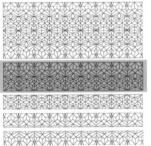
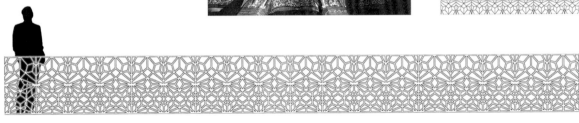

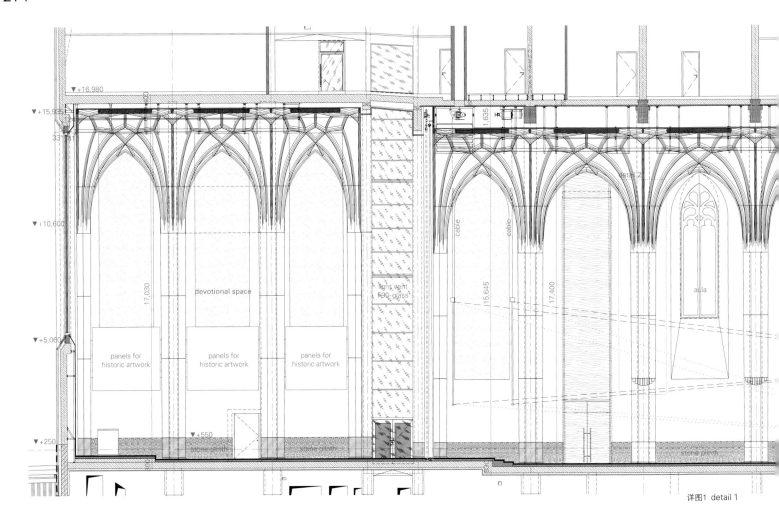

详图1 detail 1

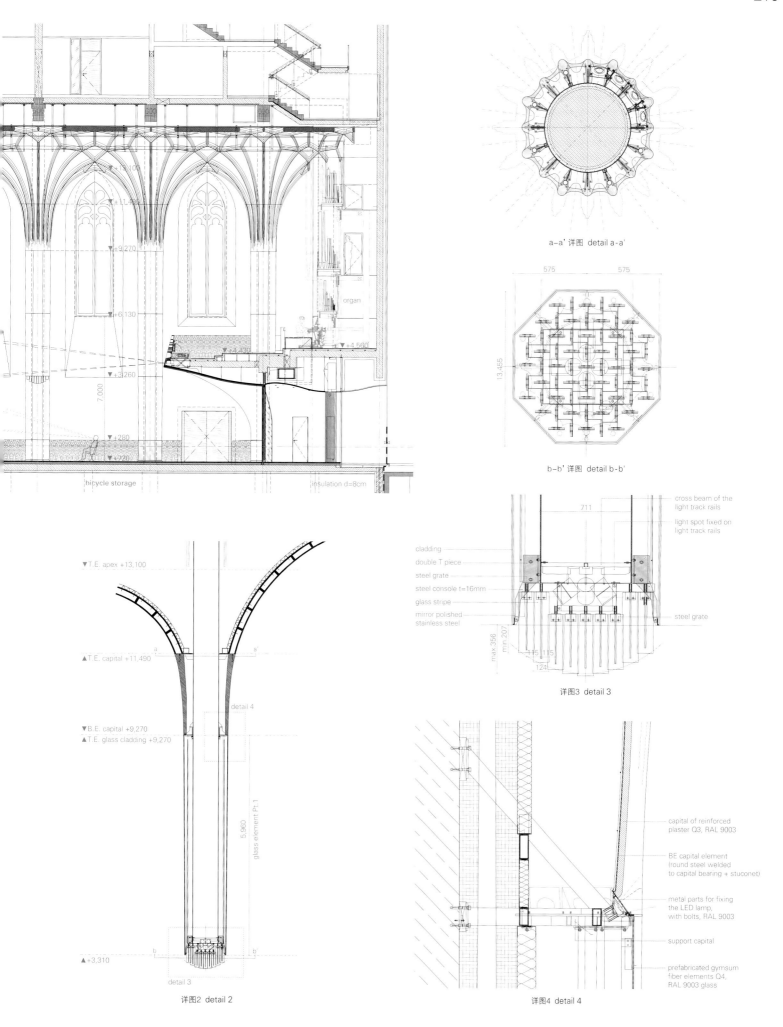

opened its doors for the students several years ago, the interior of the Audimax was officially open to the public in 2017. The Aula with Vestry (Andachtsraum), a contemporary interpretation of the former University Church, is a multi-functional space and can be used – like the original church – both for services as well as for academic ceremonies, concerts and scientific conferences.

With the original space and its cross vault as a reference, the vault construction is erected from a combination of white plaster works, crossing over into glass columns.
Among others, the auditorium space features two integrated organs and the world's highest transparent sliding doors. These doors with a height of 15.5 meters, divide the church space in two: the secular auditorium for a variety of events

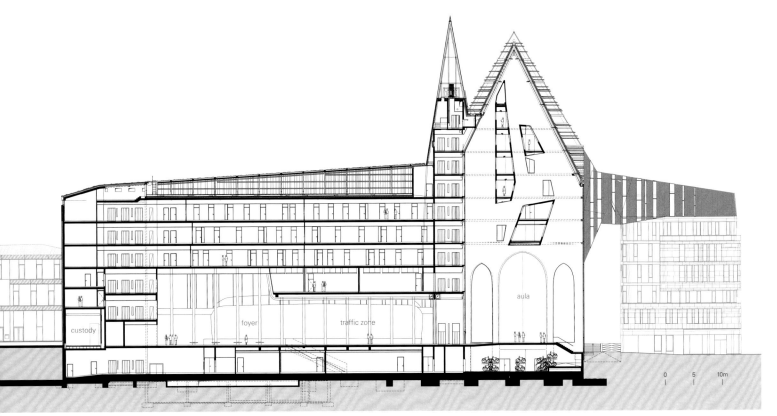

A-A' 剖面图 section A-A'

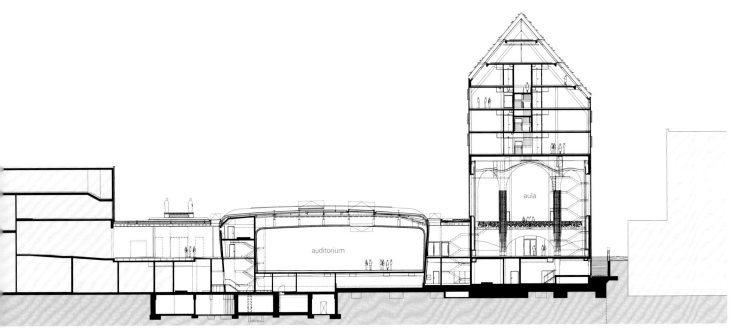

B-B' 剖面图 section B-B'

and the more contemplative space the so-called "Andachtsraum". In this part of the building the Epitaphs, which were salvaged from the ruins of the demolished church are displayed properly restored.

The impressive vaults with the glass columns – of which several column pairs do not continue to the floor to increase the functionality – ensure that the space, regardless of the glass sliding doors, can be experienced as a united whole.

With the opening of the last part of the project – the Paulinum and "Andachtsraum", the new Leipzig University located at the Augustusplatz in Leipzig, regains her dignified identity and rightfully reclaims her significance for the Leipzig, Germany and the world.

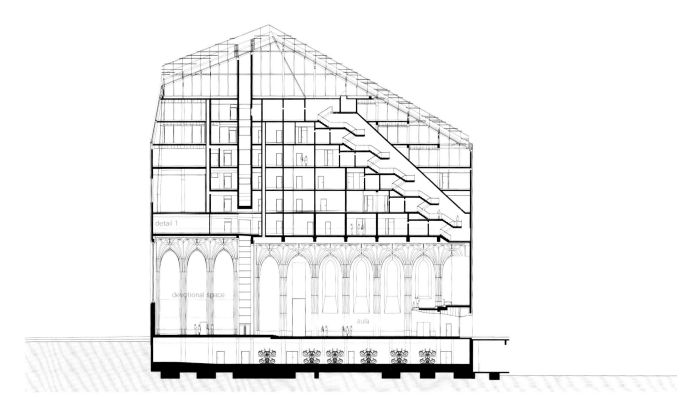

C-C' 剖面图　section C-C'

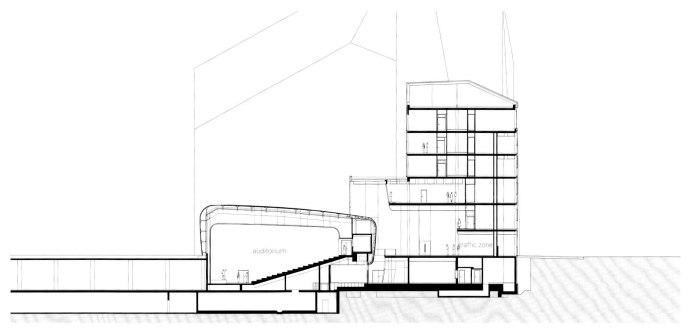

D-D' 剖面图　section D-D'

P124 **Atelier Štěpán**

Founded in Brno, Czech in 1997, its work ranges from architecture, public space and the field of design. Marek Štěpán, Principal, was born in Frýdek-Místek, Czech in 1967 and studied at the Faculty of Architecture VUT in Brno. Started Atelier Štěpán with his wife, Vanda Štěpánová, in 1997. Has worked as an assistant in the atelier of prof. Ivan Koleček on FA VUT in Brno from 1997 to 2001. Was an adviser of the Chancellor of president about architectural belongings from 2006 to 2012. Is a member of the Czech Chamber of Architects.

P194 **John McAslan + Partners**

Is based in London, with branch offices in Edinburgh and Doha. Recently opened a second office in London, the N17 Design Studio. In addition to exploring local urban regeneration projects, the new studio offers local young people apprenticeships and work placements. The practice has an established reputation both for new build projects (most recently the acclaimed Lancaster University Engineering Building) and for the imaginative regeneration of historic buildings. Notable examples of the latter include the acclaimed transformation of King's Cross Station, a Library and Student Hub at the University of Cumbria. The practice has won in excess of 115 awards including 25 RIBA Awards, three Europa Nostra Awards and the EU Prize for Cultural Heritage.

P146 **Neri&Hu Design and Research Office**

Was founded in 2004 by Lyndon Neri[right] and Rossana Hu[left]. Is based in Shanghai, China with an additional office in London, UK. Neri received his B.Arch at the UC Berkeley and a M.Arch at Harvard. Was the Director for Projects in Asia and an Associate for Michael Graves & Associates in Princeton for over 10 years. Served as an active visiting critic at the Princeton, Syracuse, Harvard GSD, and UC Berkeley. Hu received her B.Arch from the UC Berkeley and a M.Arch at Princeton. Has worked for Michael Graves & Associates, Ralph Lerner Architect, SOM, and The Architects Collaborative (TAC). Has been a guest design critic at Princeton, UC Berkeley and Syracuse.

P204 **Design Erick van Egeraat**

Erick van Egeraat is a Dutch architect, born in Amsterdam in 1956. Graduated from the Department of Architecture, TU Delft with honourable mention in 1984. Co-founded Mecanoo architects after graduation and served as partner until 1995. Soon after, established Erick van Egeraat associated architects [EA]Rwith offices in Rotterdam and Budapest. Afterwards, extended his practice to London in 1998, to Prague in 1999 and to Moscow in 2004. Restructured his company in 2009 into what is now Design Erick van Egeraat. Has led the realization of over 166 projects in more than 10 countries for 36 years. Won the RIBA Award 2007, Best Building Award 2011 & 2012 and European Property Award 2013.

P34 Plano Humano Arquitectos

Was founded in 2008 by two Portuguese architects from Lisbon, Pedro Miguel Ferreira and Helena Lucas Vieira. In 2012, they expanded their activity to Angola with several partnerships and present projects for Luanda, Huambo and Malanje. Pedro graduated in 2006 from the Lusófona University. Collaborated in 'Santos Pinheiro Arquitectos' and 'Atelier João and Luisa Sequeira'. Helena graduated in 2006 from the Faculty of Architecture, University of Lisbon, staged that same year in the department of urbanism of the Municipality of Loures. Collaborated in 'Limite Aedificandi' and 'Atelier Coreplan'. The studio was recently awarded with The American Architecture Prize 2017, and was also nominated for the Portuguese Building Awards 2017.

P14 Mario Filippetto Architetto

Is an Italian freelance architect, born in 1975. Is working in various fields including architectural design, construction supervision, structural design, design activity for fire control, and graphics applied to architecture. Studied as a free mover student in the public university of Architecture in Alicante and received a master's degree in architecture from the Polytechnic University of Milan. Has won several design competitions in Italy. Works have been published in various magazines around the world, including the Wallpaper* and the Hinge Magazine.

Anna Roos

Studied architecture at the UCT (University of Cape Town) and holds a postgraduate degree from the Bartlett School of Architecture, UCL, London. Moving to Switzerland in 2000, she worked as an architect, designing buildings in Switzerland, South Africa, Australia, and Scotland. As a freelance architectural journalist since 2007, besides C3, she also writes for A10, Ensuite Kultur Magazin, Monocle, and Swisspearl architecture magazine. Her first book, Swiss Sensibility: The Culture of Architecture in Switzerland (2017), recently published by Birkhauser Verlag, is also available in German and French editions.

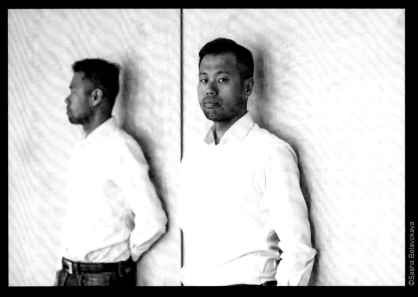

Jaap Dawson

Graduated from Cornell University with a Bachelor of Arts in English in 1971. Afterward he earned a Doctor's Degree in Education from Teachers College, Columbia University in 1979. And also received a Ingenieurs diploma(Master of Science) in Architecture from Technische Universiteit Delft in 1988. Currently he delivers a lecture of Architectural Composition in Technische Universiteit Delft and acts as architect, writer, and editor in Delft, the Netherlands since 1988.

P112 Spheron Architects

Registered Architect of ARB and a Chartered member of RIBA, Tszwai So is a British-Chinese architect. Grew up in British Hong Kong and studied architecture at the University of Hong Kong, graduating in 2003 as the recipient of Ho Fook Prize in Architecture. Later he moved to the United Kingdom and studied for a master's degree in Building History at Wolfson College, University of Cambridge. Has worked at Arc7 Design LLP from 2003 to 2011 before co-founding Spheron Architects with Samuel Bentil-Mensah in 2011. Has been a Visiting Lecturer at the Middlesex University, University of Westminster and a Guest Critic at the Central Saint Martins, Westminster University. Was named a rising star in British Architecture by the RIBA Journal in 2016 and a best UK young architect under the age of 40 by the AIA in 2017.

P182 **Fearon Hay Architects**
Was founded in 1998 in Auckland, New Zealand as a partnership between Tim Hay[p.215-upper, right] and Jeff Fearon[p.215-upper, left]. Operates with between 12 to 15 staff and delivers a diverse set of projects that include bespoke residences, restaurant & hospitality works, commercial offices, private lodges and small hotels. Projects have been or are currently being delivered in a variety of places including Sydney, Melbourne, New York, San Francisco, and Sri Lanka. Was the winner of the 2013 New Zealand Architecture Medal.

P44 **Archstudio**
Was founded by Han Wen-Qiang, an associate professor in CAFA (China Central Academy of Fine Arts). Mainly researches on contemporary architectural and interior environment based on traditional cultural background, and devotes to making space as the communication medium between people and people, people and environment. Has been featured as 2015 One of the Best Ten International Design Vanguard by Architectural Record. His works have won many awards, including LEAF (Leading European Architecture Forum) Awards Refurbishment of the Year Award, Architizer A+ Awards on Architecture + Renovation Category Jury Award and Merit Award at the Design for Asia Award.

P172 **Dans Arhitekti**
Was set up in 2004 by Miha Dešman, Eva Fišer Berlot, Rok Bogataj, Katarina Pirkmajer Dešman and Vlatka Ljubanović[from left] as partners. They all studied architecture in Ljubljana, Slovenia. Katarina and Miha studied at the Faculty of Architecture, Ljubljana and the Institute for Urbanism and Architecture in Venice. Both previously worked at the Atelje Dešman-Pirkmajer studio. Miha is a professor of history and architectural design at his alma mater. Vlatka is a teaching assistant at the Faculty of Architecture, Ljubljana. Their recent work has been nominated for the Mies Van der Rohe Award 2017 and Piranesi prize 2016. Also received Architizer A+ Award 2016 jury prize and the Plečnik award 2016 special mention.

P136 Yu Momoeda Architecture Office

Was founded in 2014 by Yu Momoeda, born in Nagasaki, Japan in 1983. He graduated from the Kyushu Institute of Design and completed his Master Course at the Yokohama Graduate School of Architecture. Has worked at Kengo Kuma & Associates and is currently teaching at the Kyushu University. The Office was named for the finalist of the AR Emerging Architecture Awards and the Leaf Awards. Also received the Gold Award of JCD International Design Awards, the Grand Award & Gold Award of the Design for Asia Awards, 2017.

P160 L.E.FT Architects

Makram el Kadi[right] was born in Beirut, Lebanon in 1974. Received his B.Arch from the American University of Beirut in 1997 and M.Arch from the Parsons School of Design in 1999. Has worked at the offices of Fumihiko Maki and Steven Holl. Taught at the GSAPP, Cornell and MIT(Aga Khan visiting Lecturer) and Yale(Louis Kahn Visiting Assistant). Is currently teaching at his alma mater. Ziad Jamaleddine[left] was born in Beirut, Lebanon in 1971. Received his B.Arch with Areen Award for excellence in design from the American University of Beirut in 1995 and M.Arch from the Harvard GSD in 1999. Has worked for Steven Holl Architects for 5 years. Taught at Cornell, PennDesign, University of Toronto, MIT(Aga Khan Program) and Yale(Louis Kahn Visiting Assistant). Is currently teaching at the GSAPP.

P90   Hariri Pontarini Architects

Founding Partner, Siamak Hariri was born in Bonn, Germany. Studied at the University of Waterloo before completing a Master of Architecture at Yale University. Has taught at the University of Toronto, and has won over 60 awards, including the Governor General's Medal in Architecture. Has established a career creating institutional and cultural projects of international acclaim, including the Schulich School of Business for York University and Casey House, both of which received the Governor General's Medal in Architecture. Also designed the Ivey School of Business at Western University, recognized with the 2016 Chicago Athenaeum International Architecture Award and the Educational Facility Design Award of Excellence by AIA.

P24   ENOTA

Was founded in 1998 in Ljubljana, Slovenia by Aljoša Dekleva, Dean Lah and Milan Tomac. Is led by Dean Lah and Milan Tomac since 2002. Dean Lah and Milan Tomac graduated from the Faculty of Architecture, University of Ljubljana in 1998. Dean Lah was born in 1971 in Maribor, Slovenia. Has been a member of executive board of Chamber of Architecture and Spatial Planning of Slovenia, member of executive board of Architects Association of Ljubljana. Milan Tomac was born in 1970 in Koper, Slovenia. From 1998 to 2001, he was Assistant Professor at his alma mater. Received The International Architecture Award, Architizer A+ Award, IOC/IAKS Award and Europe 40 under 40 Award.

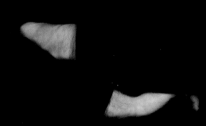

P70   Maroun Lahoud Architecte

Is a multidisciplinary platform that works in architecture at various scales as well as in object and furniture design. Maroun Lahoud is a DPLG Architect, graduated in 2004 from the Ecole d'Architecture de la Ville et des Territoires in Paris. In 2015, he moved to Beirut, Lebanon after 15 years in Paris. He teaches several Architecture Workshops at the Lebanese Academy of Fine Arts, and is currently working on various projects in Lebanon and France. For his first project, the Saint Elie Church in Brih, he received the International Archmarathon Stone Building Award in Verona in 2016, and the Lebanese Architect Award for the 'cultural building' category in Beirut in 2017.

P78   **Álvaro Siza**

Álvaro Joaquim de Melo Siza Vieira[right] was born in Matosinhos, Portugal in 1933 and graduated from the School of Fine Arts in Porto. In 1954 he opened his private practice in Porto. Taught at the in the School of Fine Arts of Porto[ESBAP], Ecole Polytechnique Fédérale de Lausanne[EPFL], and Harvard GSD as Kenzo Tange Visiting Professor. Received the Pritzker Prize in 1992, Mies van der Rohe Award in 1988, Wolf Prize for the Arts in 2001, the 2009 RIBA Royal Gold Medal, and the Golden Lion for Lifetime Achievement at the 13th International Architecture Exhibition Venice.

P78   **Carlos Castanheira**

Carlos Castanheira[left] was born in Lisbon, Portugal in 1957 and graduated from the School of Fine Arts in Porto. Lived in Amsterdam, working as an Architect and studying at the Academie voor Bouwkunst van Amsterdam. Founded Carlos Castanheira & Clara Bastai, Arquitectos Ldª with the Architect Maria Clara Bastai in 1993. Has been collaborating with Álvaro Siza in various projects since his schooldays.

P58   **Innauer-Matt Architekten**

Innauer Matt Architects, based in the Bregenzerwald, Western Austria, stand for architecture with a strong relation to an object's location, its natural surroundings and its inhabitants. Was founded by Markus Innauer[left] and Sven Matt[right] (both born in 1980) in 2012. Markus Innauer studied at the University of Applied Arts, Vienna and University of California. Is a Member of town planning board, Schruns, Austria since 2015. Sven Matt studied at the University of Innsbruck and Vienna University of Technology. Is a Board member at Vorarlberg Architecture Institute(VAI) since 2016.

© 2019大连理工大学出版社

版权所有·侵权必究

图书在版编目(CIP)数据

敬思空间：汉英对照 /（英）安娜·鲁斯等编；罗茜，于风军，王方冰译. -- 大连：大连理工大学出版社，2019.7
（建筑立场系列丛书）
ISBN 978-7-5685-2075-1

Ⅰ. ①敬… Ⅱ. ①安… ②罗… ③于… ④王… Ⅲ. ①宗教建筑－建筑设计－汉、英 Ⅳ. ①TU252

中国版本图书馆CIP数据核字(2019)第124137号

出版发行：大连理工大学出版社
（地址：大连市软件园路80号　邮编：116023）
印　　刷：上海锦良印刷厂有限公司
幅面尺寸：225mm×300mm
印　　张：14
出版时间：2019年7月第1版
印刷时间：2019年7月第1次印刷
出 版 人：金英伟
统　　筹：房　磊
责任编辑：杨　丹
封面设计：王志峰
责任校对：张昕焱
书　　号：978-7-5685-2075-1
定　　价：258.00元

发　行：0411-84708842
传　真：0411-84701466
E-mail：12282980@qq.com
URL：http://dutp.dlut.edu.cn

有印装质量问题，请与我社发行部联系更换。

建筑立场系列丛书01：
墙体设计
ISBN：978-7-5611-6353-5
定价：150.00元

建筑立场系列丛书09：
墙体与外立面
ISBN：978-7-5611-6641-3
定价：180.00元

建筑立场系列丛书17：
旧厂房的空间蜕变
ISBN：978-7-5611-7093-9
定价：180.00元

建筑立场系列丛书25：
在城市中转换
ISBN：978-7-5611-7737-2
定价：228.00元

建筑立场系列丛书33：
本土现代化
ISBN：978-7-5611-8380-9
定价：228.00元

建筑立场系列丛书41：
都市与社区
ISBN：978-7-5611-9365-5
定价：228.00元